D0073312

PUBLIC KNOWLEDGE

AN ESSAY CONCERNING THE SOCIAL DIMENSION OF SCIENCE

A very great deal more truth can become known than can be proved.

R. P. FEYNMAN

PUBLIC KNOWLEDGE

AN ESSAY CONCERNING THE SOCIAL DIMENSION OF SCIENCE

BY

J. M. ZIMAN, F.R.S.

Professor of Theoretical Physics, University of Bristol

CAMBRIDGE
AT THE UNIVERSITY PRESS
1968

Published by the Syndics of the Cambridge University Press
Bentley House, P.O. Box 92, 200 Euston Road, London, N.W. 1
American Branch: 32 East 57th Street, New York, N.Y. 10022

Library of Congress Catalogue Card Number: 68-10691

Printed in Great Britain
at the University Printing House, Cambridge
(Brooke Crutchley, University Printer)

IN MEMORY OF A GAY AND GALLANT
PHILOSOPHER AND FRIEND

NORWOOD RUSSELL HANSON

1924–1967

CONTENTS

Preface *page* ix

1 What is Science? 1

2 Science and Non-Science 13

3 Scientific Method and Scientific Argument 30

4 Education for Science 63

5 The Individual Scientist 77

6 Community and Communications 102

7 Institutions and Authorities 127

8 Summing up 143

Index 149

PREFACE

A man commonly saunters a little in turning his hand from one sort of employment to another. ADAM SMITH

Cleopatra's Nose, in the guise of Einstein's Cerebral Cortex, is now somewhat discredited as the causative agent for the events of intellectual history. Explanations in terms of movements, climates of opinion, the *Zeitgeist*, are now prescribed.

Natural Science, whose internal development for three centuries is so uniform, well-documented and relatively self-generating, is an obvious candidate for such treatment. And having noticed the *intellectual* connections between the ideas of various scholars, we must surely pass on to an investigation of the *social* relations through which those connections are established. How do scientists teach, communicate with, promote, criticize, honour, give ear to, give patronage to, one another? What is the nature of the community to which they adhere?

Surprisingly enough, when I first began to be interested in these questions, some eight years ago, I found that almost nothing had been written on the subject. There were, indeed, occasional articles in historical, sociological or professional scientific journals, but it was not until 1962 that the publication of such a collection gave the field a self-conscious identity as the *Sociology of Science*.* Since then, it has been taken over and developed by a small group of professional scholars, mainly historians and sociologists, and the 'literature' is expanding at the familiar exponential rate.

Nevertheless, this literature often lacks the authenticity of personal witness. The scholarly life is sophisticated and subtle; the would-be student of sociology approaching it with notebook and tape-recorder can no more capture its flavour than a European adult can feel his way into the psychological world of the Chinese family or, for that matter, a Confucian mandarin discover the

* Edited by B. Barber and W. Hirsch (New York: Free Press of Glencoe, 1962).

inwardness of the Catholic Church. So much is unwritten, even unspoken, and follows its own rules.

The original plan, therefore, conceived in the spring of 1964 as a joint work with Professor Jasper Rose, now at the University of California at Santa Cruz, was to describe the scholarly world in multifarious aspects, as we saw it in our own experience. That book is mapped out in some detail; but each of us has become so deeply involved in more specialized professional activities that we have not been able to push ahead with it as fast as we had hoped.

And in the course of planning the larger work, we ran into the problem of defining the whole subject matter of our investigation; to describe 'Science' we had to have some notion of its essential character. My own thoughts naturally turned in the direction of a social definition, such as I had expressed in a brief essay* some years before: this was to provide the ideological basis of our argument and the key to many explanations. But then it seemed that the unity and consistency of this somewhat novel point of view would be lost if it were deeply embedded in a grander account of academic and scientific life; as often happens even in physics, it is convenient to separate a purely theoretical, abstract and speculative line of argument from the empirical material which it exemplifies and from which it is derived. It should be much easier to indicate the significance of many of the phenomena that we shall note in the larger book if we can refer specifically to the general theory set out in the present work.

It also seems to me that several recent attempts at a general analysis of the social system of Science† have lacked just such a point of view. In order to understand how scientists interact socially, we must surely have a clear conception of what they are trying to do. Scientific research is an intensely self-conscious, deliberate and rational activity, the antithesis of those traditional, semi-rationalized, almost unconscious patterns of behaviour which are the conventional objects of anthropological and sociological

* 'Science is Social', *The Listener*, 18 Aug. 1960.

† W. O. Hagstrom, *The Scientific Community* (New York: Basic Books, 1965); N. W. Storer, *The Social System of Science* (New York: Holt, Rinehart and Winston, 1966).

scrutiny. To attempt to describe it without reference to its overt, explicit goals is to play *Hamlet* without the Prince of Denmark. I do not propose to take up every such point in these otherwise excellent and stimulating books;* but I am encouraged to publish the present work by this evidence that the views put forward in it are by no means so obvious to other people as they seem to me.

For I do not believe that the general argument of this essay is particularly revolutionary. It is implicit in the writings of a number of distinguished historians, sociologists and philosophers of science† and has not been seriously contradicted by those audiences which I have attempted to provoke by public lectures. Is it possible that this notion, too, lies immanent in the spirit of our times, only requiring to be stated to be recognized as a mere truism?

In that hope, I have endeavoured to express myself simply and clearly, as to the educated layman rather than to the connoisseur of philosophies of science. The bare bones of the argument are laid out, without all the paraphernalia of *caveats* and citations, provisos and elaborate documentation, such as would be called for in a serious work of professional scholarship. This essay does not claim to be 'scientific' in my own or anybody else's notion of the word. It is, if anything, philosophical, and amateur philosophy at that. It is written largely out of my own head, in private time spared from the pleasures of family life and from the responsibilities of academic research and teaching in a conventional scientific discipline; and makes no pretence to proper learning. The scholar who steps out of the gilded frame of his speciality is licensed to amuse the public with politico-economic *pronunciamentos*, or science-fiction romances, or as a quiz king: woe betide him if he dares to write seriously on a theme that is supposed to be the concern of other professional scholars. For this presumption may

* Some comments on Hagstrom's work are to be found in a review in *Science Progress*, **54**, 440 (1966).

† E.g. A historian, T. S. Kuhn, in *The Structure of Scientific Revolutions* (Chicago; University of Chicago Press, 1962); a sociologist, E. A. Shils, *The Torment of Secrecy* (New York: Free Press of Glencoe, 1956); a philosopher-scientist, M. Polanyi, *Personal Knowledge* (London: Routledge and Kegan Paul, 1958); a psychologist, Patrick Meredith, *Instruments of Communication* (Oxford: Pergamon Press, 1966).

mercy be granted by the historians, the sociologists and the philosophers of science who know the true facts concerning many matters upon which I must touch.

The job of a medieval jury was often to swear to a man's innocence by association with themselves: to save my bacon, let me invoke the names, and acknowledge the friendly discourse, of Richard Braithwaite, Gerd Buchdahl, Mary Hesse, Stefan Körner, Thomas Kuhn, John Lucas, John Pocock, Edward Shils, Julius Stone, Charles Weiner and Robert Young; they may not approve of the views presented here, but at least they will allow me to say that I have learnt *something* from them. And to complete my jury, I should choose, of course, my constant friend, colleague, co-editor and co-author, Jasper Rose, who provided many valuable comments on the text as it was being written, and who sees so deeply and sympathetically into the mysteries of men and affairs, Science and Art.

J. M. Z.

Bristol, December 1966

1

WHAT IS SCIENCE?

Let every man be fully persuaded in his own mind
ROM. xiv. 5

To answer the question 'What is Science?' is almost as presumptuous as to try to state the meaning of Life itself. Science has become a major part of the stock of our minds; its products are the furniture of our surroundings. We must accept it, as the good lady of the fable is said to have agreed to accept the Universe.

Yet the question is puzzling rather than mysterious. Science is very clearly a conscious artefact of mankind, with well-documented historical origins, with a definable scope and content, and with recognizable professional practitioners and exponents. The task of defining Poetry, say, whose subject matter is by common consent ineffable, must be self-defeating. Poetry has no rules, no method, no graduate schools, no logic: the bards are self-anointed and their spirit bloweth where it listeth. Science, by contrast, is rigorous, methodical, academic, logical and practical. The very facility that it gives us, of clear understanding, of seeing things sharply in focus, makes us feel that the instrument itself is very real and hard and definite. Surely we can state, in a few words, its essential nature.

It is not difficult to state the order of being to which Science belongs. It is one of the categories of the intellectual commentary that Man makes on his World. Amongst its kith and kin we would put Religion, Art, Poetry, Law, Philosophy, Technology, etc.—the familiar divisions or 'Faculties' of the Academy or the Multiversity.

At this stage I do not mean to analyse the precise relationship that exists between Science and each of these cognate modes of thought; I am merely asserting that they are on all fours with one another. It makes some sort of sense (though it may not always be stating a truth) to substitute these words for one another, in phrases like '*Science* teaches us...' or 'The Spirit of *Law* is...' or '*Technology* benefits mankind by...' or 'He is a student of *Philosophy*'. The famous 'conflict between Science and Religion' was truly a battle between combatants of the same species—between David and Goliath if you will—and not, say, between the Philistine army and a Dryad, or between a point of order and a postage stamp.

Science is obviously like Religion, Law, Philosophy, etc. in being a more or less coherent set of ideas. In its own technical language, Science is information; it does not act directly on the body; it speaks to the mind. Religion and Poetry, we may concede, speak also to the emotions, and the statements of Art can seldom be written or expressed verbally—but they all belong in the non-material realm.

But in what ways are these forms of knowledge *unlike* one another? What are the special attributes of Science? What is the criterion for drawing lines of demarcation about it, to distinguish it from Philosophy, or from Technology, or from Poetry?

This question has long been debated. Famous books have been devoted to it. It has been the theme of whole schools of philosophy. To give an account of all the answers, with all their variations, would require a history of Western thought. It is a daunting subject. Nevertheless, the types of definition with which we are familiar can be stated crudely.

Science is the Mastery of Man's Environment. This is, I think, the vulgar conception. It identifies Science with its products. It points to penicillin or to an artificial satellite and tells us of all the wonderful further powers that man will soon acquire by the same agency.

This definition enshrines two separate errors. In the first place it confounds Science with Technology. It puts all its emphasis on the applications of scientific knowledge and gives no hint as to the intellectual procedures by which that knowledge may be successfully obtained. It does not really discriminate between Science and

Magic, and gives us no reason for studies such as Cosmology and Pure Mathematics, which seem entirely remote from practical use.

It also confuses ideas with things. Penicillin is not Science, any more than a cathedral is Religion or a witness box is Law. The material manifestations and powers of Science, however beneficial, awe-inspiring, monstrous, or beautiful, are not even symbolic; they belong in a different logical realm, just as a building is not equivalent to or symbolic of the architect's blueprints. A meal is not the same thing as a recipe.

Science is the Study of the Material World. This sort of definition is also very familiar in popular thought. It derives, I guess, from the great debate between Science and Religion, whose outcome was a treaty of partition in which Religion was left with the realm of the Spirit whilst Science was allowed full sway in the territory of Matter.

Now it is true that one of the aims of Science is to provide us with a Philosophy of Nature, and it is also true that many questions of a moral or spiritual kind cannot be answered at all within a scientific framework. But the dichotomy between Matter and Spirit is an obsolete philosophical notion which does not stand up very well to careful critical analysis. If we stick to this definition we may end up in a circular argument in which Matter is only recognizable as the subject matter of Science. Even then, we shall have stretched the meaning of words a long way in order to accommodate Psychology, or Sociology, within the Scientific stable.

This definition would also exclude Pure Mathematics. Surely this is wrong. Mathematical thinking is so deeply entangled with the physical sciences that one cannot draw a line between them. Modern mathematicians think of themselves as exploring the logical consequences (the 'theorems') of different sets of hypotheses or 'axioms', and do not claim absolute truth, in a material sense, for their results. Theoretical physicists and applied mathematicians try to confine their explorations to systems of hypotheses that they believe to reflect properties of the 'real' world, but they often have no licence for this belief. It would be absurd to have to say that Newton's *Principia*, and all the work that was built upon it, was not now Science, just because we now suppose that the inverse square law of gravitation is not perfectly true in an Einsteinian universe. I suspect that the exclusion of the 'Queen of the Sciences' from her throne is a relic of some ancient academic arrangement, such as the combination of classical literary studies with mathematics in the Cambridge Tripos, and has no better justification than that Euclid and Archimedes wrote in Greek.

Science is the Experimental Method. The recognition of the importance of experiment was the key event in the history of Science. The Baconian thesis was sound; we can often do no better today than to follow it.

Yet this definition is incomplete in several respects. It arbitrarily excludes pure mathematics, and needs to be supplemented to take cognisance of those perfectly respectable sciences such as Astronomy or Geology where we can only observe the consequences of events and circumstances over which we have no control. It also fails to give due credit to the strong theoretical and logical sinews that are needed to hold the results of experiments and observations together and give them force. Scientists do not in fact work in the way that operationalists suggest; they tend to look for, and find, in Nature little more than they believe to be there, and yet they construct airier theoretical systems than their actual observations warrant. Experiment distinguishes Science from the older, more speculative ways to knowledge but it does not fully characterize the scientific method.

Science arrives at Truth by logical inferences from empirical observations. This is the standard type of definition favoured by most serious philosophers. It is usually based upon the principle of induction—that what has been seen to happen a great many times is almost sure to happen invariably and may be treated as a basic fact or Law upon which a firm structure of theory can be erected.

There is no doubt that this is the official philosophy by which most practical scientists work. From it one can deduce a number of practical procedures, such as the testing of theory by 'predictions' of the results of future observations, and their subsequent confirmation. The importance of speculative thinking is recognized, provided that it is curbed by conformity to facts. There is no restriction of a metaphysical kind upon the subject matter of Science, except that it must be amenable to observations and inference.

But the attempt to make these principles logically watertight does not seem to have succeeded. What may be called the positivist programme, which would assign the label 'True' to statements that satisfy these criteria, is plausible but not finally compelling. Many philosophers have now sadly come to the conclusion that there is no ultimate procedure which will wring the last drops of uncertainty from what scientists call their knowledge.

And although working scientists would probably state that this is the Rule of their Order, and the only safe principle upon which their discoveries may be based, they do not always obey it in practice. We often find complex theories—quite good theories—that really depend on very few observations. It is extraordinary, for example, how long and complicated the chains of inference are in the physics of elementary particles; a few clicks per month in an enormous assembly of glass tubes, magnetic fields, scintillator fluids and electronic circuits becomes a new 'particle', which in its turn provokes a flurry of theoretical papers and ingenious interpretations. I do not mean to say that the physicists are not correct; but no one can say that all the possible alternative schemes of explanation are carefully checked by innumerable experiments before the discovery is acclaimed and becomes part of the scientific canon. There is far more faith, and reliance upon personal experience and intellectual authority, than the official doctrine will allow.

A simple way of putting it is that the logico-inductive scheme does not leave enough room for genuine scientific error. It is too black and white. Our experience, both as individual scientists and historically, is that we only arrive at partial and incomplete truths;

we never achieve the precision and finality that seem required by the definition. Thus, nothing we do in the laboratory or study is 'really' scientific, however honestly we may aspire to the ideal. Surely, it is going too far to have to say, for example, that it was 'unscientific' to continue to believe in Newtonian dynamics as soon as it had been observed and calculated that the rotation of the perihelion of Mercury did not conform to its predictions.

This summary of the various conceptions of science obviously fails to do justice to the vast and subtle literature on the subject. If I have empasized the objections to each point of view, this is merely to indicate that none of the definitions is entirely satisfactory. Most practising scientists, and most people generally, take up one or other of the attitudes that I have sketched, according to the degree of their intellectual sophistication—but without fervour. One can be zealous for Science, and a splendidly successful research worker, without pretending to a clear and certain notion of what Science really is. In practice it does not seem to matter.

Perhaps this is healthy. A deep interest in theology is not welcome in the average churchgoer, and the ordinary taxpayer should not really concern himself about the nature of sovereignty or the merits of bicameral legislatures. Even though Church and State depend, in the end, upon such abstract matters, we may reasonably leave them to the experts if all goes smoothly. The average scientist will say that he knows from experience and common sense what he is doing, and so long as he is not striking too deeply into the foundations of knowledge he is content to leave the highly technical discussion of the nature of Science to those self-appointed authorities the Philosophers of Science. A rough and ready conventional wisdom will see him through.

Yet in a way this neglect of—even scorn for—the Philosophy of Science by professional scientists is strange. They are, after all, engaged in a very difficult, rather abstract, highly intellectual activity and need all the guidance they can from general theory. We may agree that the general principles may not in practice be very helpful, but we might have thought that at least they would be taught to young scientists in training, just as medical students are taught Physiology and budding administrators were once encouraged to acquaint themselves with Plato's *Republic*. When the student graduates and goes into a laboratory, how will he know what to do to make scientific discoveries if he has not been taught the distinction between a scientific theory and a non-scientific one? Making all allowances for the initial prejudice of scientists against speculative philosophy, and for the outmoded assumption that certain general ideas would communicate themselves to the educated and cultured man without specific instruction, I find this an odd and significant phenomenon.

The fact is that scientific investigation, as distinct from the theoretical *content* of any given branch of science, is a practical art. It is not learnt out of books, but by imitation and experience. Research workers are trained by apprenticeship, by working for their Ph.Ds under the supervision of more experienced scholars, not by attending courses in the metaphysics of physics. The graduate student is given his 'problem': 'You might have a look at the effect of pressure on the band structure of the III–V compounds; I don't think it has been done yet, and it would be interesting to see whether it fits into the pseudopotential theory'. Then, with considerable help, encouragement and criticism, he sets up his apparatus, makes his measurements, performs his calculations, etc. and in due course writes a thesis and is accounted a qualified professional. But notice that he will not at any time have been made to study formal logic, nor will he be expected to defend his thesis in a step by step deductive procedure. His examiners may ask him why he had made some particular assertion in the course of his argument, or they may enquire as to the reliability of some particular measurement. They may even ask him to assess the value of the 'contribution' he has made to the subject as a whole. But they will not ask him to give any opinion as to whether Physics is ultimately *true*, or whether he is justified now in believing in an external world, or in what sense a theory is verified by the observation of favourable instances. The examiners will assume that the candidate shares with them the common language and principles of their discipline. No scientist really doubts that theories are verified by observation, any more than a Common Law judge hesitates to rule that hearsay evidence is inadmissible.

What one finds in practice is that scientific argument, written or spoken, is not very complex or logically precise. The terms and concepts that are used may be extremely subtle and technical, but they are put together in quite simple logical forms, with expressed or implied relations as the machinery of deduction. It is very seldom that one uses the more sophisticated types of proof used in Mathematics, such as asserting a proposition by proving that its negation implies a contradiction. Of course actual mathematical or numerical analysis of data may carry the deduction through many steps, but the symbolic machinery of algebra and the electronic circuits of the computer are then relied on to keep the argument straight.* In my own experience, one more often detects elementary *non sequiturs* in the verbal reasoning than actual mathematical mistakes in the calculations that accompany them. This is not said to disparage the intellectual powers of scientists; I mean simply that the reasoning used in scientific papers is not very different from what we should use in an everyday careful discussion of an everyday problem.

This is a point to which we shall return in a later chapter. It is made here to emphasize the inadequacy of the 'logico-inductive' metaphysic of Science. How can this be correct, when few scientists are interested in or understand it, and none ever uses it explicitly in his work? But then if Science is distinguished from other intellectual disciplines neither by a particular style or argument nor by a definable subject matter, what is it?

The answer proposed in this essay is suggested by its title: *Science is Public Knowledge*. This is, of course, a very cryptic definition, with almost the suggestion of a play upon words.† What I

* This point I owe to Professor Körner.

† There is also, unfortunately, the hint of an antithesis to *Personal Knowledge*, the title of Polanyi's book to which I have already referred. No antagonism is meant. Polanyi goes a long way along the path I follow, and is one of the few writers on Science who have seen the social relations between scientists as a key factor in its nature.

mean is something along the following lines. Science is not merely *published* knowledge or information. Anyone may make an observation, or conceive a hypothesis, and, if he has the financial means, get it printed and distributed for other persons to read. Scientific knowledge is more than this. Its facts and theories must survive a period of critical study and testing by other competent and disinterested individuals, and must have been found so persuasive that they are almost universally accepted. The objective of Science is not just to acquire information nor to utter all non-contradictory notions; its goal is a *consensus* of rational opinion over the widest possible field.

In a sense, this is so obvious and well-known that it scarcely needs saying. Most educated and informed people agree that Science is true, and therefore impossible to gainsay. But I assert my definition much more positively; this is the basic principle upon which Science is founded. It is not a subsidiary consequence of the 'Scientific Method'; it *is* the scientific method itself.

The defect of the conventional philosophical approach to Science is that it considers only two terms in the equation. The scientist is seen as an individual, pursuing a somewhat one-sided dialogue with taciturn Nature. *He* observes phenomena, notices regularities, arrives at generalizations, deduces consequences, etc. and eventually, Hey Presto! a Law of Nature springs into being. But it is not like that at all. The scientific enterprise is corporate. It is not merely, in Newton's incomparable phrase, that one stands on the shoulders of giants, and hence can see a little farther. Every scientist sees through his own eyes—and also through the eyes of his predecessors and colleagues. It is never one individual that goes through all the steps in the logico-inductive chain; it is a group of individuals, dividing their labour but continuously and jealously checking each other's contributions. The cliché of scientific prose betrays itself 'Hence *we* arrive at the conclusion that...' The audience to which scientific publications are addressed is not passive; by its cheering or booing, its bouquets or brickbats, it actively controls the substance of the communications that it receives.

In other words, scientific research is a social activity. Technology, Art and Religion are perhaps possible for Robinson Crusoe, but Law and Science are not. To understand the nature of Science, we must look at the way in which scientists behave towards one another, how they are organized and how information passes between them. The young scientist does not study formal logic, but he learns by imitation and experience a number of conventions that embody strong social relationships. In the language of Sociology, he learns to play his *role* in a system by which knowledge is acquired, sifted and eventually made public property.

It has, of course, long been recognized that Science is peculiar in its origins to the civilization of Western Europe. The question of the social basis of Science, and its relations to other organizations and institutions of our way of life, is much debated. Is it a consequence of the 'Bourgeois Revolution', or of Protestantism—or what? Does it exist despite the Church and the Universities, or because of them? Why did China, with its immense technological and intellectual resources, not develop the same system? What should be the status of the scientific worker in an advanced society; should he be a paid employee, with a prescribed field of study, or an aristocratic dilettante? How should decisions be taken about expenditure on research? And so on.

These problems, profoundly sociological, historical and political though they may be, are not quite what I have in mind. Only too often the element in the argument that gets the least analysis is the actual institution about which the whole discussion hinges—scientific activity itself. To give a contemporary example, there is much talk nowadays about the importance of creating more effective systems for storing and indexing scientific literature, so that every scientist can very quickly become aware of the relevant work of every other scientist in his field. This recognizes that publication is important, but the discussion usually betrays an absence of careful thought about the part that conventional systems of scientific communication play in sifting and sorting the material that they handle. Or again, the problem of why Greek Science never finally took off from its brilliant taxying runs is discussed in terms of, say, the aristocratic citizen despising the servile labour of practical experiment, when it might have been due to the absence of just such a communications system between scholars as was provided in the Renaissance by alphabetic printing. The internal sociological analysis of Science itself is a necessary preliminary to the study of the Sociology of Knowledge in the secular world.

The present essay cannot pretend to deal with all such questions. The 'Science of Science' is a vast topic, with many aspects. The very core of so many difficulties is suggested by my present argument—that Science stands in the region where the intellectual, the psychological and the sociological coordinate axes intersect. It is knowledge, therefore intellectual, conceptual and abstract. It is inevitably created by individual men and women, and therefore has a strong psychological aspect. It is public, and therefore moulded and determined by the social relations between individuals. To keep all these aspects in view simultaneously, and to appreciate their hidden connections, is not at all easy.

It has been put to me that one should in fact distinguish carefully between Science as a body of knowledge, Science as what scientists do and Science as a social institution. This is precisely the sort of distinction that one must *not* make; in the language of geometry, a solid object cannot be reconstructed from its projections upon the separate cartesian planes. By assigning the intellectual aspects of Science to the professional philosophers we make of it an arid exercise in logic; by allowing the psychologists to take possession of the personal dimension we overemphasize the mysteries of 'creativity' at the expense of rationality and the critical power of well-ordered argument; if the social aspects are handed over to the sociologists, we get a description of research as an N-person game, with prestige points for stakes and priority claims as trumps. The problem has been to discover a unifying principle for Science in all its aspects. The recognition that scientific knowledge must be public and *consensible* (to coin a necessary word) allows one to trace out the complex inner relationships between its various facets. Before one can distinguish and discuss separately the philosophical, psychological or sociological dimension of Science, one must somehow have succeeded in characterizing it as a whole.*

In an ordinary work of Science one does well not to dwell too long on the hypothesis that is being tested, trying to define and describe it in advance of reporting the results of the experiments or calculations that are supposed to verify or negate it. The results themselves indicate the nature of the hypothesis, its scope and limitations. The present essay is organized in the same manner. Having sketched a point of view in this chapter, I propose to turn the discussion to a number of particular topics that I think can be better understood when seen from this new angle. To give a semblance of order to the argument, the various subjects have been arranged according to whether they are primarily *intellectual*—as, for example, some attempt to discriminate between scientific and non-scientific disciplines; *psychological*—e.g. the role of education, the significance of scientific creativity; *sociological*—the structure of the scientific community and the institutions by which it maintains scientific standards and procedures. Beyond this classification, the succession of topics is likely to be pretty haphazard; or, as the good lady said, 'How do I know what I think until I have heard what I have to say?'

The subject is indeed endless. As pointed out in the Preface, the present brief essay is meant only as an exposition of a general theory, which will be applied to a variety of more specific instances in a larger work. The topics discussed here are chosen, therefore, solely to exemplify the main argument, and are not meant to comprehend the whole field. In many cases, also, the discussion has been kept abstract and schematic, to avoid great marshlands of detail. The reader is begged, once more, to forgive the inaccuracies and imprecisions inevitable in such an account, and to concentrate his critical attention upon the validity of the general principle and its power of explaining how things really are.

* 'Hence a true philosophy of science must be a philosophy of scientists and laboratories as well as one of waves, particles and symbols.' Patrick Meredith in *Instruments of Communication*, p. 40.

2
SCIENCE AND NON-SCIENCE

Opinion in good men is but knowledge in the making
MILTON

In this chapter Science will be considered mainly in its intellectual aspects, as a system of ideas, as a compilation of abstract knowledge. The first question to be answered has already been posed in the introductory chapter: what distinguishes Science from its sister 'Faculties'—Law, Philosophy, Technology, etc.? The argument is that Science is unique in striving for, and insisting on, a consensus.

Take Law, for example. We all feel that legal thought is quite different from scientific thought—but what is the basis of this intuition? There are many ways in which legal argument is very close to Science. There is undoubtedly an attempt to make every judgement follow logically on statutes and precedents. Every lawyer seeks to clarify a path of implications through successive stages to validate his case. The judge reasons it out, on the basis of universal principles of equity, in the effort to arrive at a decision that will command the assent of all just and learned men.

The kinship of Law with the mathematical sciences is emphasized by the interesting suggestion that legal decisions might be arrived at automatically by a computer, into which all the conditions and precedents of the case would be fed and a purely mechanical process of logical reduction would produce exactly the correct judgement.* Although perhaps the idea is somewhat fanciful, if this procedure were technically feasible it would provide decisions that could not but command the assent of all lawyers—just as a table of values of a mathematical function printed out by a computer commands the assent of all mathematicians. To the

* I am indebted to Professor Julius Stone for sending me his fascinating critical essay on this subject.

extent, therefore, that the Law is strictly logical, it can be made 'scientific'.

Again, in the concept of 'evidence' there is close similarity. This is too primary and basic an idea to be defined readily, but, roughly speaking, it means 'any information that is relevant to a disputed hypothesis'. In Science, as in Law, we are almost always dealing with theories that are disputable, and that can only be challenged by an appeal to evidence for and against them. It is the duty of scientists, as of lawyers, to bring out this evidence, on both sides, to the full.

In the end, the case may hang upon some very minor item of information—was the man who got off the 3.57 at Little Puddlecome on Monday, 27 May, wearing a black hat? A scientific theory also may be validated by some tiny fact—for example, the almost imperceptible changes in the orbit of the planet Mercury. The question of the *credibility* of evidence can become very important. We may find everyone in full agreement that, if a fact is as stated by a witness, it has vital logical implications for the hypothesis under consideration; yet the court may be completely undecided as to whether this evidence is true or not. The existence of honest error has to be allowed for. This sort of thing happens in Science too, though it does not usually get remembered in the conventional histories. For example, many scientists will recall the interest that was aroused by the publication of evidence for organic compounds in meteorites—probably an erroneous interpretation of a complex observation, but of the most profound significance if it proved to be true. In such cases there may even be questions about the relative reliability, in general, of two different observers—an assessment, perhaps, based upon their scientific standing and expert authority—just as the relative veracity of conflicting witnesses may become the key issue in a legal case.

But, of course, in Science, when the evidence is conflicting, we withhold our assent or dissent, and do the experiment again. This cannot be done in legal disputes, which must be terminated yea or nay. If we are forced to a premature opinion on a scientific question, we are bound to give the Scottish verdict *Not Proven*, or say that the jury have disagreed, and a new trial is needed. In Criminal Law, where the case for the prosecution must be proved up to the hilt, or the accused acquitted, this is well enough; but Civil Law demands a decision, however difficult the case.

The Law is thus unscientific because it *must* decide upon matters which are not at all amenable to a consensus of opinion. Indeed, legal argument is concerned with the conflict between various principles, statutes, precedents, etc.; if there were not an area of uncertainty and contradiction, then there would be no need to go to law about it and get the verdict of the learned judge. In Science, too, we are necessarily interested in those questions that are not automatically resolved by the known 'Laws of Nature' (the analogy here with man-made Laws is only of historical interest) but we agree to work and wait until we can arrive at an interpretation or explanation that is satisfactory to all parties.

There are other elements in the Law that are quite outside science—normative principles and moral issues that underlie any notion of justice. As is so often said, Science cannot tell us what *ought* to be done; it can only chart the consequences of what *might* be done.

Normative and moral principles cannot, by definition, be embraced in a consensus; to assert that one *ought* to do so and so is to admit that some people, at least, will not freely recognize the absolute necessity of not doing otherwise. Legal principles and norms are neither eternal nor universal; they are attached to the local, ephemeral situation of this country here and now; their arbitrariness can never be mended by any amount of further logical manipulation. Thus, there are components of legal argument that are necessarily refractory to the achievement of free and general agreement and these quite clearly discriminate between Law and Science as academic Faculties.

To the ordinary Natural Scientist this discussion may perhaps have seemed quite unnecessary—Law, he would say, is a man-made set of social conventions, whilst Science deals with material, objective, eternal verities. But to the Social Scientist this distinction is by no means so clear. He may, for example, find it impossible to disentangle such legal concepts as personal responsibility from his scientific understanding of the power of social determination in a pattern of delinquency. The criterion of consensibility might temper some of the scientific arrogance of the expert witness—'Would *every* criminologist agree with you on this point, Dr X?'—whilst at the same time throwing the full weight of personal decision and responsibility upon the judge, who should never be allowed to shelter behind the cruel and mechanical absolutes of 'Legal Science'. The intellectual authority of Science is such that it must not be wielded incautiously or irresponsibly.

At first sight, one would not suppose that much need be said about what distinguishes Science from those disciplines and activities that belong to the Arts and Humanities—Literature, Music, Fine Arts, etc. Our modern view of Poetry, say, is that it is an expression of a private personal opinion. By his skill the poet may strike unsuspected chords of emotion in a vast number of other men, but this is not necessarily his major intention. A poem that is immediately acceptable and agreeable to everyone must be banal in the extreme.

But, of course Arts dons do not write Poetry: they write about it. Literary and artistic critics do sometimes pretend that their judgements are so convincing that it is wilful to oppose them. An imperious temper demands that we accept their every utterance of interpretation and valuation. Fortunately, we have the right of dissent and if our heart and mind carry us along a different path we have no need to be frightened by their shrill cries of contempt.

The point here is that there are genuine differences of taste and feeling, just as there are genuine differences of moral principle. At the back of our definition of Science itself is the assumption that men are free to express their true feelings; without this condition, the notion of a consensus loses meaning. Under a dictatorship we might be constrained to pay lip service to a uniform standard of style or taste, but this is the death of criticism.

There are, of course, periods of 'classicism' and 'academicism' when some style or technique is overwhelmingly praised and practised, but no one supposes that this is in obedience to the commands of absolute necessity. The attempts of the stupider sort of academic critic to rationalize the taste of his age by rules of 'harmony' or 'dramatic unity' are invariably by-products of the fashion whose dominance he is seeking to justify, not its determining factors. No sooner are such rules formulated than a great artist cannot resist the temptation to break them, and a new fashion sweeps the land. By their very nature, the Arts are not consensible, and hence are quite distinct from Science as I conceive it.

Science is not immune from fashion—a sure sign of its socio-psychological nature. We shall return to some of the symptoms of this disease later. But what, abstractly, *is* fashion? It means doing what other people do for no better reason than that that is what is done. If everyone were to follow only fashionable lines of thought, there would be a false impression of a consensus; the inhibition of the critical imagination by such a conformist sentiment is the antithesis of the scientific attitude. It is also, of course, another way of death for true Poetry and Art.

But the products and producers of Literature, Art and Music may be studied in more factual aspects than for their emotional or spiritual message. For example, they are the outcome of, or participants in, historical events.

The place of *History* in this analysis is very significant; it seems to be truly one of the borderlands marching between scientific and non-scientific pursuits. Suppose that we are investigating such a problem as the date and place of birth of a writer or statesman. We search in libraries and other collections of material documents for written evidence. From various oblique references we might build up an argument in favour of some particular hypothesis—an argument to persuade our colleagues by its invincible logic that no other interpretation is tenable. This procedure seems quite as scientific as the research of a palaeontologist, who might reconstruct the anatomy of an extinct animal by piecing together fragments of fossil bone. Our aim is the same—to make a thoroughly convincing case which no reasonable person can refute. If, unfortunately, we cannot find sufficient evidence to clinch the case, we do not cling to our hypothesis and abuse our opponents for not accepting it; we quietly concede that the matter is uncertain, and return once more to the search. On such material points, the mood of historical scholarship is perfectly scientific.

The other mood in History is much more akin to Literature or Theology; it is the attempt to understand human history imaginatively and to 'explain' it. Having ascertained the 'facts', the historian tries to uncover the hidden motives and forces at work, just as the scientist goes behind the phenomena to the laws of their being.

The trouble is that the complex events of history can seldom be explained convincingly in the language of elementary cause and effect. To ascribe the English Civil War, for example, to the 'Rise of the Gentry' may be a brilliant and fruitful hypothesis, but it is almost impossible to prove. Even though one may feel that this is the essence of the matter, and though one may marshal factual evidence forcefully in its favour, the case can be no more than circumstantial and hedged with vagueness and provisos. It will go into the canon of interesting historical theories, but experience tells us that it will not, as would a valid scientific theory, be so generally acceptable as to eliminate all competitors.

The rule in Science is not to attempt explanations of such complex phenomena at all, or at least to postpone this enterprise until many much simpler and more direct problems have been solved. In Science we deliberately restrict our attention to questions whose answers are capable of being agreed upon. Imagination in the search for such problems is essential, but speculation is always kept rigidly under control. Even in such disciplines as Cosmology, where it sometimes seems as if a new theory of the Universe is promulgated each week, the range of discussion is limited quite narrowly to model systems whose mathematical properties are calculable and can be critically assessed by other scholars.

History does not impose such restrictions upon its pronouncements. It is felt, quite naturally, that the larger questions, although more difficult, are very important and must be discussed, even if they cannot be answered with precision. To restrict oneself to decidable propositions would be to miss the lessons that the strange sad story has for mankind. A history of 'facts', of dates and kings and queens, although acceptable to the consensus, would be banal and trivial. In other words, History also has to provide other spiritual values, and to satisfy other normative principles, than scientific accuracy.

There are, of course, historians who have claimed universal 'scientific' validity for their larger schemes and 'Laws'. It is not inconceivable that historical events do follow discernable patterns, and that there are, indeed, hidden forces—the class struggle, say, or the Protestant ethic—which largely determine the outcome of human affairs. It would not be necessary for such a theory to be absolute and mathematically rigorous for it to acquire scientific validity, any more than the proof that smoking causes lung cancer requires every smoker to die at the age of 50. It is not inherently absurd to search for historical laws, any more than it was absurd, 200 years ago, to search for the laws governing smallpox. Seemingly haphazard events often turn out to have their pattern, and to be capable of rational explanation.

All I am saying is that no substantial general principles of historical explanation have yet won universal acceptance. There have been fashionable doctrines, and dogmas backed by naked force, but never the sort of consensus of free and well-informed scholars that we ordinarily find in the Natural Sciences. Many historians assert that historical events are the outcome of such a variety of chance causes that they could never be subsumed to simpler, more general laws. Others say that the number of instances of exactly similar situations is always too small to provide sufficient statistical evidence to support an abstract theoretical analysis.

Whatever their reasons, historians do not agree on the general theoretical foundations or methodology of their studies. Instead of establishing, by mutual criticism and tacit cooperation, a limited common basis of acceptable theory, from which to build upwards and outwards, they often feel bound to set up antagonistic 'schools' of interpretation, like so many independent walled cities.

They are not to be blamed for such behaviour; it only shows that this is a field where a scientific consensus is not the main objective. If you insisted that historians should work more closely together, they would object that the knowledge that they have in common is too dull, too trivial, too distant from the interesting problems of History, to circumscribe the thought of a serious scholar. To write about the Civil War without asking why the whole extraordinary thing happened is to compose a mere chronicle. For that reason, much of historical scholarship is not essentially scientific.

It would be wrong, on the other hand, only to give the approving label 'Science' to the new techniques of historical research derived from the physical and biological sciences and technologies—carbon dating, aerial photography, demographic statistics, chemical analysis of ink and parchment. Such techniques are often powerful, but they are not more 'scientific' than the traditional scholarly exercises of editing texts, verifying references and making rational deductions from the written words of documents. There is no reason at all why marks on paper in comprehensible language should be treated as inherently less evidential than the pointer readings of instruments or the print-out from a computer. In German the word *Wissenschaft*, which we translate as *Science*, includes quite generally all the branches of scholarship, including literary and historical studies.

To maintain, therefore, an impassable divide between Science and the Humanities is to perpetrate a gross misunderstanding that springs in the British case, solely from a peculiarity of educational curricula. The Story, the Arts, the Poetry of Mankind are worthy both of spiritual contemplation and scholarly study, whether by laymen in general education or by experts as their life career. In many aspects this study is perfectly akin to the scientific study of electrons, molecules, cells, organisms or social systems: consensible knowledge may be acquired whether as isolated facts or as generally valid explanations. But to confine oneself, in education as in scholarship, to such aspects would impoverish the imagination, and even restrict the scope of possible further advance. Without general concepts as a guide—however uncertain, personal and provisional

—we simply could not see any larger patterns in the picture. Historical and literary scholarship cannot therefore pretend to be scientific through and through, but that does not prevent their making progress towards a closer definition of the truth. In the end, bold speculative generalization and unverifiable psychological insight may go further in establishing a convincing narrative than a rigid insistence on precise minutiae.

It scarcely needs to be said that *Religion*, as we nowadays study and practise it, is also quite distinct from Science. This seems so obvious in our enlightened age that one wonders how there could have been any conflict and confusion between them. But was not Religion primitive Science—the corpus of generally accepted public knowledge? Should we not see Science as growing out of, and eventually severing itself from, this parent body—or perhaps as a process of differentiation and specialization within the unity of the medieval *Summum*? Just because many religious beliefs are now seen to be wrong, it does not follow that they were not seriously, freely and rationally accepted in their time. Conventional science too can be wrong at times.

Let me give an example. In the late eighteenth and early nineteenth century, prehistoric remains were found that we now see as pointing to the great antiquity of Man. But many scholars stood out against this interpretation because it did not square with the Biblical chronology of the past. Is it fair to treat this as a conflict between scientific rationality and religious prejudice? Would it not be more just to say that a widely accepted theory was being ousted by a better one as new evidence came to light?

The point is that this debate was open and free. The participants on one side may have been blinkered by their upbringing, but their beliefs were honestly held and rationally maintained. They may often have used poor arguments to defend their case—but they did not call in the secular arm or the secret police. In the end, they lost; and since then the appeal to Divine Scripture has ceased to be an acceptable element in a scientific discussion.

What I am arguing is that there is a progressive improvement in the techniques and criteria of such discussion, and that the use of abstract theological principles was once respectable but is now discredited, just as the absolute justification of Euclidean geometry from the Parallelism Axiom is now discredited. The 'Scientific Revolution' of the seventeenth century is not a complete break with the past. The idea of presenting a rational non-contradictory account of the universe is perhaps a legacy of Greece, but it is very strong in medieval Philosophy and Theology. It may be that the very existence of a dogmatic system of metaphysics, implying a rational order of things and fiercely debated in detail, was the prerequisite for the development of an alternative system, using some of the same logical techniques but based upon different principles and more extensive evidence.* The doctrinaire consensus of the Church may have been prolonged beyond its acceptability to free men by the power of the Holy Office, but it had originally provided an example of a generally agreed picture of the world. These are subtle and deep questions which I am not competent to discuss, but I wonder whether the failure of Science to grow in China and India was due as much to the general doctrinal permissiveness of their religious systems as to any other cause. Toleration of deviation, and the lack of a very sharp tradition of logical debate may have made the very idea of a consensus of opinion on the Philosophy of Nature as absurd to them as the idea of absolute agreement on ethical principles would be to us.

The relationship between Science and *Philosophy* is altogether more complex and confused. In a sense, all of modern science is the Philosophy of Nature, as distinct from, say, Moral or Political Philosophy. But this terminology is somewhat old-fashioned, and we try to make a distinction between Physics and Metaphysics, between the Philosophy *of* Science and Philosophy *as* Science. Some philosophers attempt to limit themselves to statements as precise and verifiable as those of scientists, and confine their arguments to the rigid categories of symbolic logic. The consensus

* This point is made in *Science in the Modern World* by A. N. Whitehead (New York: Macmillan, 1931).

criterion would be acceptable to them, for they would hold that by a continuous process of analysis and criticism they would make progress towards creating a generally agreed set of principles governing the use of words and the establishment of valid truths. Others hold that such a hope can never be realized and that by limiting philosophical discourse in this way they would only allow themselves to make trivial statements, however unexceptionable. For this school of Philosophy it is important to be free to comment on grander topics, even though such comments will only reveal the variety and contradictory character of the views of different philosophers.

As with History we can only say that if Philosophy is what academic philosphers write in their books, then some of it is not very different from Science. But generally the motivation is non-scientific, by our definition, and the multiplicity of viewpoints indicates that there is no dominant urge to find maximum regions of agreement. Whatever their claims, the proponents of 'scientific' philosophical systems do not convince the majority of their colleagues that theirs is the only way to truth.

* * *

Let us now consider *Technology*—Engineering, Medicine, etc. For the multitude, Science is almost synonymous with its applications, whereas scientists themselves are very careful to stress the distinction between 'pure' knowledge, studied 'for its own sake', and technological knowledge applied to human ends.

The trouble is that this distinction is very difficult to make in practice. Suppose, for example, that we are researching on the phenomenon of 'fatigue' in metals. We are almost forced into the position of saying that on Monday, Wednesday and Friday we are just honest seekers after truth, adding to our understanding of the natural world, etc., whilst on Tuesday, Thursday and Saturday we are practical chaps trying to stop aeroplanes from falling to pieces, advancing the material welfare of mankind and so on. Or we may have to make snobbish distinctions between Box, a pure scientist working in a University, and Cox, a technologist, doing the same research but employed by an aircraft manufacturer. There was once a time when Science was academic and useless and Technology was a practical art, but now they are so interfused that one is not surprised that the multitude cannot tell them apart.

Here again, a definition in terms of the scientific consensus can be really effective. The technologist has to fulfil a need; he must provide the means to do a definite job—bridge this river, cure this disease, make better beer. He must do the best he can with the knowledge available. That knowledge is almost always inadequate for him to calculate the ideal solution to his problem—and he cannot wait while all the research is done to obtain it. The bridge must be built this year; the patient must be saved today; the brewery will go bankrupt if its product is not improved.

So there will be a large element of the incalculable, of sheer art, in what he does. A different engineer would come out with quite a different design; a different doctor would prescribe quite different treatment. These might be better or worse, in their results—but nobody quite knows. Each situation is so complex, and has so many unassessable factors, that the only sensible policy for the client is to choose his engineer or doctor carefully and then rely upon his skill and experience. To look for a solution acceptable to all the professional experts is a familiar recipe for disaster—'Design by Committee'.

The technologist's prime responsibility is towards his employer, his customer or his patient, not to his professional peers. His task is to solve the problem in hand, not to address himself to the opinions of the other experts. If his proposed solution is successful, then it may well establish a lead, and eventually add to the 'Science' of his Technology; but that should not be in his mind at the outset.

What we find, of course, is that a corpus of generally accepted principles develops in every technical field. Modern Technology is deliberately scientific, in that there is continuous formal study and empirical investigation of aspects of technique, in addition to the mere accumulation of experience from successfully accomplished tasks. The aim of such research is not to solve immediate specific problems, but to acquire knowledge for the use of the experts in their professional work. It is directed, therefore, at the mind of the profession, as a potential contribution to the consensus opinion. This sort of work is thus genuinely scientific, however trivial and limited its scope may be.

The abstract distinction here being made between a 'scientific Technology' and 'technological Science' has its psychological counterpart. It is a commonplace in the literature on the Management of Industrial Research that applied scientists often suffer from divided loyalties. On the one hand, they owe their living to the company that employs them, and that expects its return in the profitable solution of immediate problems. On the other hand, they give their intellectual allegiance to their scientific profession—to Colloid Chemistry, or Applied Mathematics, or whatever it is—where they look for scholarly recognition. Although the rewards for 'technological' work are greater and more direct, they very often prefer to stick to their 'scientific' research.

This preference seems almost incomprehensible to management experts, because they fail to see that the scientific loyalty is not just towards a prestigious professional group but to an ideology. The young scientist is trained to make contributions to public knowledge. All the habits and practices of his years of apprenticeship emphasize the importance of making them convincing, and thus making them part of the common pool. Being a successful scientist is not just winning prizes; it is having other scientists cite your work. To give this up is worse than losing caste; it is to give up one's faith and be made to worship foreign idols.

Nevertheless, one must agree that Science and Technology are now so intimately mingled that the distinction can become rather pedantic. Take, for example, a typical Consumers Association report on a motor car. Some of the tests, such as the measure of petrol consumption, may be perfectly scientific in that their validity would be universally acceptable. Other tests, such as whether the springing was comfortable, would not satisfy this criterion, although it would be one of the important skills of the designer to attend to just such 'subjective' and 'qualitative' features. For this reason, to say that a car has been 'scientifically designed' is merely to assert that it has been well designed by competent engineers. Yet an account, by the designer, of the rationale behind various technical features of the model could rank as a serious contribution to the Science of Automobile Engineering by adding to the body of agreed principles at the basis of that mysterious art.

All that I can claim is that these distinctions, although subtle and perhaps pedantic, are not entirely arbitrary or unreal. We do not need to look far ahead to some conceivable remote application of the knowledge in question, nor do we need to examine the hidden, perhaps unconscious, motives of those who produce it. We do not need to decide whether some particular laboratory is 'technological' or 'scientific', and then attach the appropriate label to its products. The criterion is in the work itself, in the form in which it is presented, and the audience to which it is addressed.

What are we to make of the so-called 'Social Sciences' in the light of this discussion? It is obvious that such a subject as Politics is very close to History and to Philosophy in its goals and achievements; to stick to ascertainable public 'facts' is to limit the discourse to the banal. To give this discipline the name of Political Science is unfortunate; it offers more than it can deliver and debases its ethical message.

On the other hand, *Economics* is a very technological subject; the experts are always being asked to diagnose the ills of the nation and to propose specific cures, long before they have sufficient scientific understanding to make a valid analysis. Yet the totally quantitative material medium—money—allows of convincing proofs, statistical or algebraic, of precise hypotheses, so that a body of agreed principles is gradually emerging. Leading economists may debate in public, and seem to be at loggerheads, but behind the scenes they teach much the same things to their students. It is typical of the tacit cooperation between scholars in a scientific subject that American and Soviet economists respect and learn from each other's work on Input-Output Analysis, however much they may disagree on more speculative issues of general social policy.

These are the neighbours of the new discipline of *Sociology*. which is an attempt to escape from the 'unscientific' traditions of History and Politics, and to make the study of social systems, and of man in society, at least as scientific as Economics.

That is the reason why so much sociological research is by questionnaire and statistical analysis; the aim is to provide the necessary factual basis for firmly scientific theories. To the extent that observations of this sort are verifiable by repetition, and capable of being made quite convincing to a critical public, this attitude is sound. But the intractability of the subject must be reckoned with. Vast quantities of information do not add up to much serious knowledge without theories to give it meaning. Moreover, even to accept 'facts' of this sort may imply the acceptance of dubious hidden theories. Suppose, for example, that our car-testing organization decided to assess the comfort of various models by asking a hundred people to give their views, and reported that car A 'rated' 87 per cent and car B only 63 per cent. This is objective information, which might well be 'verifiable'. But there is the implication that it 'measures' something—'comfort'—which may not exist at all. This is a crude case, but sociological research is full of more subtle examples of the same difficulty.

Some sociologists have taken quite a different line. They deal in abstract categories, which they manipulate logically into various hypothetical relations in a sort of formal calculus. This approach also strives towards the creation of a consensus, in that the structure of the argument can be purged of contradictions and hence made unexceptionable and theoretically acceptable. But without much more rigorous connection of these abstractions with real systems and actual phenomena it is vacuous knowledge, without the power to persuade us that thus and thus is the world of men.

Nevertheless, Sociology is often genuinely scientific in spirit, although it has turned out to be an exceedingly difficult science whose positive achievements do not always match the effort expended on it. The 'methodological problem' has not been surmounted; there is not yet a reliable procedure for building up interesting hypotheses that can be made sufficiently plausible to a sufficient number of other scholars by well devised observations, experiments or rational deductions. It was the sort of problem facing Physics before Galileo began seriously to apply mathematical reasoning and numerical measurement to the subject. The ideal of a consensus is there, but the intellectual techniques by which it might be created and enlarged seem elusive.

* * *

This survey of the Faculties has necessarily been brief and schematic. Why should we even want to decide whether a particular discipline is scientific or not? The answer is, simply, that, *when it is available*, scientific knowledge is more reliable, on the whole, than non-scientific. When there are conflicts of authority, when Sociology tells us to go one way and History another, we need to weigh their respective claims to validity. Our general argument here is that in a discipline where there is a scientific consensus the amount of *certain* knowledge may be limited, but it will be honestly labelled: 'Trust your neck to this', or 'This ladder was built by a famous scholar, but no one else has been able to climb it'.

In the end, the best way to decide whether a particular body of knowledge is scientific or not is often to study the attitudes of its professional practitioners to one another's work. A sure symptom of non-science is personal abuse and intolerance of the views of one scholar by another. The existence of irreconcilable 'schools' of thought is familiar in such academic realms as Theology, Philosophy, Literature and History. When we find them in a 'scientific' discipline, we should be on our guard.

This is the reason why for example we should be very suspicious of the claims of Psychoanalysis. The history of this subject is a continuous series of bitter conflicts between persons, schools and theories. Freud himself had the most honest and sincere desire to create a thoroughly respectable scientific discipline, but for some reason he failed to understand this key point—the need to move slowly forward, step by step, from a basis of generally accepted ideas. Perhaps the struggle to get anyone to listen at all was too bitter, or perhaps his mind was too active and impatient to endure continuous critical assessment of each new theory or interpretation. Whatever the reason, the mood of Psychoanalysis in its formative period was antagonistic to the covert cooperative spirit of true Science. Its clinical successes were only of technological significance, and did not scientifically validate the theories on which they were said to be based.

I have given this example, not out of prejudice against psychoanalytic ideas (one or other of the contending schools may well be right: we shall see) but to show that the principle of the consensus is a powerful criterion, with something definite to say on this vexed topic. To some people the words 'scientific' and 'unscientific' have come to mean no more than 'true' and 'false', or 'rational' and 'irrational'. In this chapter I have tried to show, by reference to other organized bodies of knowledge, that this usage is quite improper, and grossly unfair to those scholars who seek rationality and truth in bolder ways than by microscopic dissection of minutiae.

3

SCIENTIFIC METHOD AND SCIENTIFIC ARGUMENT

In Greece, for the first time, appeared a handy means by which one could put the logical screws upon somebody, so that he could not come out without admitting either that he knew nothing, or that this and nothing else was the truth, the eternal *truth that never would vanish as the doings of the blind men vanish*

MAX WEBER

At this point I find myself in a difficult position. Having put forward a new approach to a general philosophy of Science, I should give serious consideration to the vast literature that already exists on this major scholarly theme. I should read my way into the subject, discussing seriously such questions as the validity of the Principle of Induction, or the logical status and prescriptions for Laws, Models, etc. I should indicate how such and such confirms or conforms to the consensus principle, how so and so's objections are to be met, and so on. To shirk this duty is to duck away from the best-armed and trained of all possible critics.

Would this really be so helpful?

In the only book that I have found where the consensus idea is seriously discussed, Norman Campbell* derives the whole conventional apparatus of 'the Scientific Method' from the following definition: 'Science is the study of those judgements concerning which universal agreement can be obtained'. From this sentence, which is obviously very close to the position taken in this book, Campbell arrives very convincingly at a common-sense version of the logico-inductive scheme.

Yet I am sceptical as to whether this or any other essentially positivist view of science can be salvaged in this way. Campbell interprets the definition rather more narrowly and abstractly than

* *What is Science* (London: Methuen, 1921).

I would. Physics is his paradigm of a science; he does not quite admit the extreme difficulties of establishing quantitative laws in other fields. Nor does he discuss the sociological connotations of the consensus notion; his scientists are still very individual and personal seekers after truth.

These are just the restrictions that make 'the Philosophy of Science' nowadays so arid and repulsive. To read the latest symposium volume on this topic is to be reminded of the Talmud, or of the theological disputes of Byzantium. It is not now a field where the amateur philosopher may gently wander and pick a few nosegays. It is fiercely professional and technical and almost meaningless to the ordinary working scientist. All too often it becomes an exercise in 'jobbing backwards': it tells us how we ought to have derived our result if only we had known the answer before we began. Or its applications have been to those parts of Science that are already complete, agreed and established—as if we were continually applying the principles of Moral Philosophy to instances out of the Bible or of Greek Tragedy.

This is doubly unfortunate: the divorce of Science from Philosophy impoverishes both disciplines. A philosophical sense is needed in Science—a nose for a generalization or a logical contradiction, for false analogies and for instructive parallels—just as experience in scientific research may correct a tendency to fruitless abstraction in a philosopher.

In this chapter, therefore, I shall try to heal the breach by talking semi-philosophically about some of the intellectual procedures of scientific investigation. What are the implications of the consensus ideal for the techniques of research and scientific argument? What is the meaning of experiment, theory, discovery, prediction, logic, in this light?

I shall try to show that the familiar 'method' of Science, whatever its logical status and epistemological virtue, also has tremendous *rhetorical* power. If applied correctly, it has overwhelming persuasive force. We use these techniques, consciously or unconsciously, not only to unravel the secrets of Nature for ourselves but also to reveal them, in full daylight, to our colleagues. Even if we cannot completely justify and validate the normal procedures of Science, they are not inexplicable as devices for the construction of a free consensus.

The word *rhetoric* here may seem out of place: it carries hints of bolstering a bad argument by appeals to the emotions rather than to the intellect. But this surely is the only word we may use, once we have dethroned positivism, and challenged the absolutism of 'scientific' proof. It implies an appropriate degree of scepticism and doubt concerning any alleged scientific discovery, without suggesting that we should treat the whole matter as a fraud.* Why, in fact, do we *believe* a good scientific argument, whether or not we can give a complete logical justification for that belief?

* * *

Consider the idea of *experiment*. There are several strands of significance twisted together in this main cable of the scientific method.

In one sense, an experiment is merely a reproducible observation. Its essence is that the experience recorded shall be under prescribed circumstances, so that it may be repeated by the original experimenter, or by anyone else. A single observer, however honest, may be deceived; it is essential, for scientific accuracy, that others too should see what happens. By the repetition of the experiment, they validate its results, and incidentally convince themselves thoroughly on the same point. Thus, experimental evidence is public knowledge, *par excellence*, with the power of carrying complete conviction.

The rhetorical mechanism here is quite simple: 'Seeing is Believing'. The poor fools who refused to look through Galileo's telescope knew what they were about; if they did see the moons of Jupiter, then they would be *forced* to believe what they did not wish to believe; therefore it was wiser not to look.

There is a deeper explanation. Modern psychology, especially

* The rehabilitation of rhetoric as the theory of sound argument is suggested in the works of Perelman and Meredith, to which the reader is referred for much wise comment.

such work as that of Jean Piaget on children, has shown that the human mind does not contain a ready-made model of reality against which to measure its experiences. Such a model, and the procedures of observation and perception that further enrich it, is gradually created by these experiences themselves, by their similarities and contrasts, by their intrinsic pattern. Once a child has become self-conscious, through language, this process is strongly influenced, if not dominated, by communication with other minds. We are all entirely conditioned to accept as absolute and real the public view of things that we can share with other humans. An experimental observation that one may reproduce for oneself is thus utterly convincing; this is the primitive mechanism by which all the rest of our 'real world'—all those things in which we totally believe—has been constructed. It takes more courage to reject an experimental result than to accept it utterly.

Experiment is more than mere *reproducible* observation; it implies that the circumstances are new—that it is original. Here we encroach upon a topic that is of great fascination, but that we must deliberately skirt—the psychology of invention. Experiment as the tool of discovery—that is the heart of the skill of the active scientist. How to devise an experiment that will confirm our theory or lay bare a new realm of unsuspected phenomena—that is what we ask ourselves, night and day, as working scientists.

But there is a world of difference between our own experiment and somebody else's. This is a point that needs the very strongest emphasis. Science is not made by our own experiments alone; it is made by everybody's. Our experiment may be brilliant, and utterly convincing to *us*—but it must carry its weight with other people before it is acceptable.

The conventional description of scientific research* concentrates wholly on the significance of one's *own* experiments and observations. But the material of science is everyone's experiments—anyone's experiments. The transformation between the personal and the public is not trivial, in practice or in principle.

* E.g. in W. I. Beveridge's *The Art of Scientific Investigation* (New York: Norton; London: Heinemann, 1950).

For example, many experiments will never be published at all; we regard them as 'faulty' or 'inconclusive' or 'negative'.* On the other hand, we may repeat some parts of the work several times, with new apparatus or in new ways, not to convince ourselves further of the validity of our conclusions but to 'tidy up any loose ends' or to 'round off the argument'. The work as published is no mere chronicle of the research as it took place; it is a much more contrived document, with its logical teeth brushed and its observational trouser seams sharply creased. It is written in a curiously artificial 'impersonal' style, deliberately flat and unemotive, as from one calculating machine to another. The experiment is not now something that really occurred to me, the author; it is what always takes place, in principle, under the ideal circumstances set out in the paper.

The machinery of scientific communication will be discussed in a later chapter. The point here is that we gain our knowledge of science by reading and thinking about other people's experiments, which we seldom actually repeat for ourselves. This would be almost impossible in practice, and is in fact discouraged except in crucial cases; merely to rediscover the results of already published investigations is considered quite unenterprising and trivial. We mainly rely upon the honesty and competence of the original observers. In other words, the critical role of each scientist with respect to the experimental work of his colleagues is somewhat abrogated. But it still exists in principle; there is always the likelihood that an experiment will be repeated; always the danger that a different result may be obtained. What we do, therefore, to guard against this danger is to play critic against ourselves. The abstract, impersonal formulation of our 'paper' is an attempt to anticipate all objections. We set out the experiment, and report its results, in the ideal form that we are seeking to have accepted as if it already occupied an established position in the consensus.

This mechanism, by which observations are made 'objective'

* What is the status of experimental results that are not published? It seems to me that this might raise the same sort of difficulties for some philosophies of science as the question of the soul of a still-born child does for some theologies!

and criticism is internalized, is of the greatest importance in science, and will be discussed at length later. What I am getting at here is that the report of an experiment is a very long way indeed from that direct and strenuous wrestling with brute Nature that the individual research worker experiences in his own laboratory.

When we say that scientific knowledge is empirical, or is firmly based upon empirical evidence, we do not mean that each scientist has seen with his own eyes all the wonders in which he believes. We mean that there exists a collection of reports of observations made by reliable witnesses and set out according to certain conventions, clearly and unambiguously, with due attention to possible sources of error. These reports are not diaries or journals, telling us exactly what occurred in a particular laboratory on a particular day. They give, rather, a carefully edited version of such events, and inform us what ought to happen if you try to repeat the experiment yourself under the prescribed conditions. The guarantee of the validity of the information is not that the recipe has been tried ten thousand times successfully, by ten thousand different people in ten thousand different laboratories (which is the sort of warrant demanded for an 'inductive generalization') but that the original experimenters are well-trained, skilful and honest, and know that their assertions *could* be verified by anyone willing to take so much trouble. I am not saying that these reports are in any way unreliable (although the case of 'Piltdown Man' indicates the havoc that can be caused by a single false item in science) but simply that by becoming part of the stock of public knowledge they have become hearsay evidence, second-hand information, far removed from the direct experience of any one of us.*

And of course a really *good* experiment, a really novel and exciting one, is connected, in the mind of the experimenter, with the proof of some novel and exciting hypothesis. His communication of the experiment to his colleagues is not merely an exposition of the peculiar events that occurred when he put a piece of litmus paper in the solution; it is an attempt to show that the world behaves as he has conceived it. After the private moment of illumi-

* Polanyi makes this point particularly strongly.

nation, there must come the public demonstration, the deliberate process of persuasion. That is why I say that a good experiment is a powerful piece of rhetoric; it has the ability to persuade the most obdurate and sceptical mind to accept a new idea; it makes a positive contribution to public knowledge.

Experiment bridges the gulf between the empirical and the theoretical. The basic substance of *theory* is Logic, or reasoning. It is sometimes said in books on scientific method that science 'does not depend on reason'. Literally, this is nonsense, but the implication is that scientific knowledge is based as far as possible on observation and experiment, rather than on ratiocination.

Nevertheless, the Laws of Logic, and all statements that can be derived rigorously from them, are part of the 'public domain' of knowledge. The most trivial deduction from experiment is likely to appeal to principles such as 'if s implies p and p implies q then s implies q', or 'Socrates is a man; all men are mortal; therefore Socrates is mortal'. There may be interesting problems for the logician in discussing the precise formal relations between these laws, and there are even more interesting problems for the psychologist in the study of how each one of us comes to take them utterly for granted, but the fact is that they are well understood, and entirely accepted throughout the scientific world. There are classes of human discourse, such as Zen poetry, which do not conform to these principles but these are distinctly non-scientific.

With Logic, we accept Mathematics. Again, the precise nature of the kinship between Logic and Mathematics is a matter of debate; but nobody nowadays doubts that most of Mathematics is quite sound, in the sense that its theorems can be made to follow quite rigorously from its axioms. This also is 'public knowledge', and genuinely scientific. It would be idle, for example, to dispute the validity of the hundreds of equations and formulae in a book such as Whittaker and Watson's *Modern Analysis* (assuming that these contain no technical errors) except in some refined philosophical sense of no practical significance. These equations are persuasive to any mind that can learn the rules by which they are constructed; they are a consequence of the very definitions of their terms. Far from 'not depending on reason' in Science, we try to put as much scientific argument as we can into mathematical form, which is reasoning made rigorous by symbolic operations.

Our definition of Science therefore automatically includes all of 'Pure' Mathematics, as well as its applications. As I have argued in chapter 1, to attempt to draw a distinction between mathematical results proved only for hypothetically postulated axioms, and results derived from axioms that are thought to be 'really true', leads to nonsense, or at least to fluctuating boundary lines that cannot be drawn with any certainty at any moment. We sometimes say of a theoretical physicist that he 'only does mathematics' in the sense that his results may be of only abstract formal validity, without direct relevance to observable phenomena, but we do not drum him out of the scientific community for heresy. And the object of work in Pure Mathematics is to persuade other mathematicians, thereby adding to the general body of accepted mathematical knowledge.

Within the empirical sciences, however, the role of Mathematics and Logic is not so definite as within its own realm. The point is that a 'theory' is more than a succession of formal mathematical deductions that can be rigorously proved; it is an argument that has to make contact with the observable, and that contains postulates and models that cannot be deduced from elsewhere. It is these elements, these relations, that cause so much trouble in the attempt to reduce scientific procedures to a logical method.

My view is that this part of scientific theory, although it may be discussed and analysed, is primarily extralogical—but not thereby irrational. A scientific argument is an attempt to make some 'theory' extremely plausible and convincing, but cannot be strictly deductive. This is, of course, a central point in the argument of this book, for otherwise we should seek an alternative definition of Science which would use logical 'method' to validate all true scientific results. But it is my impression, from all that I have read in the Philosophy of Science, that such a 'method' does not exist. Some of the reasons for this belief have been indicated already; the reader is referred to the works of Polanyi, Körner and Perelman* for critical discussions in its favour.

What are the features then of a conventional scientific argument, that make it so convincing? Why do we believe in scientific knowledge, even if we do not believe that it is *necessarily*, in a logically rigorous manner, true? The strongest argument, surely, is that a theory provides a logical ordering, a pattern, for observations. It seems to run something like this: we observe four vertical posts projecting from below the fringe of a curtain. They move in unison. Then somebody says 'There is a chair there'—and suddenly it make sense.

This is a deliberately naïve account of one of the most characteristic and complex features of the human mind, but I think it is the prototype of even the most sophisticated of scientific discoveries and insights. As professional scientists, we learn to manipulate mental concepts much more elaborate and abstruse than tables and chairs, but that does not make the basic psychological mechanism more, or less, mysterious. This is obviously one of the major human capabilities that makes Science possible, and those who have attempted to conjoin the intellectual and psychological aspects of Science† have quite properly made it their main theme.

Yet the power to conceptualize is far wider than Science in its scope. In the language of Euclid, it is a *necessary* but not *sufficient* condition for the growth of a corpus of Public Knowledge. I propose, therefore, not to discuss it further, but to take it for granted, just as I do not discuss the validity of symbolic logic nor the possibility of communicating from mind to mind. Let us recall our purpose of trying to see Science in *all* its dimensions, and avoid the Circean delights of purely 'philosophical' argument.

One further condition may be necessary for consensible knowledge—a common metaphysic. Those who participate in the consensus must already share many beliefs.

* M. Polanyi, *Personal Knowledge*; S. Körner, *Experience and Theory* (London: Routledge and Kegan Paul, 1966); Ch. Perelman, *The Idea of Justice and the Problem of Argument* (London: Routledge and Kegan Paul, 1963).

† Especially Polanyi and Meredith.

It is obvious, for example, that they must agree that such a consensus is possible and desirable. An entirely arbitrary and ineffable Universe, of the type professed, for example, by Kim's Red Lama, would simply not permit such unity of opinion about itself. As I have already suggested, the totalitarianism of Christian Theology may have provided just such a conviction of the rationality and comprehensibility of the world—a firm belief that there really was something that men might investigate with profit.

Must the metaphysic of Science go further? Must it, for example, deny all spiritual, immanent or transcendental phenomena and values? This does not seem essential. The familiar fact that many excellent scientists have been devoutly and conventionally religious has often been a stumbling block for those who hold that Science implies the truth of their particular brand of materialism, agnosticism or atheism. From our point of view, it is no more difficult to suppose that scientists cooperate to reveal the marvellous handiwork of the Lord than to imagine that their task is to unmask the errors of biblical or other authority.

As I see it, scientists need not hold that *all* knowledge is potentially reducible to science, or that *only* scientific knowledge is credible. But they must believe that *some* truths can be discovered by the scientific method—i.e. by publication, criticism and acceptance by a free community—and that these are basically reliable. It may have been necessary historically for that community to share a more specific metaphysic, such as belief in atoms, cause and effect, the primacy of the experimental approach, God as mathematician and so on; but such particular doctrines do not seem essential to Science in principle, and many of them have been shed as the consensus itself has expanded and engulfed them. In other words, Science creates its own internal metaphysic, which need not—could not—be pre-existent to itself.

The key to the reconciliation of Science with Religion, Poetry and other forms of wisdom is the power to discriminate personally between what is consensible and public in the scientific sense, and what is not. The virtue called for is no more than that toleration of diversity in others, and modest scepticism concerning one's own cherished ideas, that is required at every stage in research itself. One does indeed need the power to compartmentalize one's mind—not with watertight bulkheads, but into a number of interconnecting chambers labelled by the strength and character of the evidence on which one's beliefs are based. As I shall try to show, the room labelled Science is not sharply cut off from its neighbours but opens in every direction to less specific, less 'certain' apartments of the House of Intellect.

Science as a whole, therefore, is credible by our and its existence. The problem of its validity is no more, or less, deep than the elementary philosophical problem of whether the external world exists. All we say is that if such a world does exist, this is what it is like; and if it does not exist then this is what it seems to be like. I do not see that Science calls for an epistemological analysis distinct from the general problem of knowledge. An experiment consists in pushing one leg of the chair and seeing the others move. It is confirmation, or a further extension of the pattern. It is another piece of the jigsaw puzzle that clicks into place. The wonder is not the 'existence' of ribonucleic acids or neutrinos, but of tables and chairs, knives and forks, men, women and dogs.

Let us return, then, to our investigation of the rhetorical power of the scientific method. Scientists continually refer to 'agreement between theory and experiment'. This is what makes a scientific paper convincing: a pattern of theory is shown to imply features which are confirmed by observation. If the theory is well ordered logically—i.e. is capable of public acceptance as reasonably self-consistent and non-contradictory—and if the observation is conducted 'objectively'—i.e. in a publicly verifiable and repeatable way—then the concordance can immediately be accepted as public knowledge, and hence scientifically reputable.

Reference is often made to the *predictions* of a theory. This suggests that certain phenomena, as yet unobserved, are consequences of the theoretical model—for example, that light will be bent in passing close to the sun, as a consequence of the principles of General Relativity. The confirmation of such a prediction is, of course, a day of rejoicing in the laboratory, and of congratulation for the author of the theory.

Why should this be so? Logically speaking, there is no difference between prediction and retrodiction. The pattern of theory is usually hinged to certain established points—for example, the values of various numerical parameters—from which a number of other experimentally observable facts, such as the values of other experimentally measurable quantities, are deduced. It does not matter, surely, whether those other facts have already been determined; the logical fitting and the 'verification' of the theory is just as good as if we then sat down to measure them for the first time.

This is true in Logic, but not in Rhetoric. What we say to ourselves is: if he did not already know the answer to the hypothetical experiment, then he could not have 'cooked' his theory to agree with it; therefore it is much more convincing. The element of predictability in good science is just what makes us believe in it so strongly.

Perhaps this is why many philosophers of science regard Physics as the paradigm of science. Physics may be defined as the study of systems that can be reduced to mathematical terms. In Physics, when we find a complicated object that cannot be described in simple geometry or by a few equations, we knock it into smaller pieces, or we purify and recrystallize it, or we cool it down, or we heat it up, until it becomes amenable to mathematical analysis. Theories in Physics must therefore always be mathematical, and hence allow of strong logical deductions, often of a numerical kind. The confirmation of these deductions, especially very close agreement between theoretically predicted and experimentally observed quantities, is extraordinarily convincing of the validity of that theory, once one has grasped its nature. It is also possible to test the model behind the theory at many points, so that it becomes inconceivable to disbelieve it. Nobody, for example, who has studied elementary quantum theory, and understood the way that it accounts for innumerable quantitative features of atoms, nuclei, molecules, etc. so that every symbol in Schrödinger's wave equation has, so to speak, been confirmed ten thousand times by whole books of tables of atomic data, is likely to deny its essential validity. At the 'frontiers', where scientific knowledge is being laboriously acquired, Physics is as crude and uncertain as any other discipline, but it has the power to subdue, chain down, discipline and civilize new theories, once they are reasonably established, so that they become as 'absolute', within a few years, as the most venerable principles of Newton or Faraday or Maxwell. In other sciences, where this reduction to the mathematically amenable would quite destroy the subject matter of the investigation—a complex molecule, a cell, a human being—such predictive power, such close numerical agreement between theory and experiment, is almost always unattainable, so that every theory remains a little tentative, and an experiment can only be plausible rather than utterly convincing. The theoretical models are more complex, and not to be derived logically from a few abstract postulates; one always has the feeling that quite a different theoretical model might give just the same experimental conclusion.

To my mind, this does not demonstrate that Physics is inherently more 'scientific' than Neurophysiology or Anthropology. The choice between mathematically precise general statements about the bricks, and the imprecise but more interesting statements that one would like to make about the building or the city, cannot be avoided; for all the power of modern computing devices, it is inconceivable that we should ever be able to reduce a protein, a cell, or a man to a mathematical equation with predictable properties. It is unfortunate that so many students of the History and Philosophy of Science should take Physics as their ideal, and suppose that all good scientific arguments must be of the mathematical/physical kind.* I agree with Meredith† when he says 'Physics is unique among the sciences', not because it is inherently concerned only with the most general properties of the material world, but because that generality and 'elementality' are forced upon it by the demands of mathematical definability. For all the other sciences, the

* It amuses me to try to apply such arguments to the first piece of research I ever did, which was an attempt to discover how much grass a pig normally ate in a day!

† *Instruments of Communication.*

cost of such analytical precision would be too high, and other, less persuasive forms of argument become necessary.

Nevertheless, *number* and *quantity*, which are very public and objective qualities, may be defined far outside the realms of the physical sciences. Kelvin's dictum, which would make Science co-extensive with measurement, goes too far, but it is still true that numerical and quantitative comparisons are of the greatest importance in fields where the full power of mathematical theory is quite inapplicable. To give an extreme example, the counting of names in Parish Registers* can tell us something about village life in the seventeenth century that we could never feel sure of from purely anecdotal evidence.

It is important to note, however, that such data only play the role of Ordinal Numbers; they allow comparisons of greater or less, more or fewer, but do not turn up as the solutions of algebraic equations or expressed through formulae involving other theoretical magnitudes.

This must be emphasized, because all too often the attempt is made to manipulate such numbers statistically. How far may we, so to speak, try to coerce the consensus by such methods? What is the power of a statistical demonstration of a scientific theory?

The problem here is not the classical philosophical question of the logical status of Probability Theory, which is a perfectly respectable branch of Mathematics, but the very difficult topic of Statistical Inference. In the physical sciences we learn something of the theory of errors, and soon accept the fact that even an experiment to confirm Boyle's Law will not give points lying on a mathematical straight line. Some degree of imprecision does not vitiate the theory. An experiment that succeeds 999 times is not utterly falsified by failure on the thousandth trial. The fact that all empirical observations are necessarily subject to statistical fluctuations is a key point in the attack on absolute scientific positivism, but it is not of much practical importance when very high precision is in fact attainable.

* P. Laslett *The World We Have Lost* (London: Methuen, 1965).

The question is: how much statistical error can we tolerate? From Probability Theory one can deduce formulae for the 'significance' of a result—for example, that there is only a chance of 1 in 20 that the apparently superior effectiveness of drug A over drug B could have occurred by accident. Is this good enough for 'scientific' proof?

As I see it, results of such low significance are not strong enough to be incorporated in the scientific canon. When one is searching for evidence, following up hunches, looking for the slightest clue, then such data are valuable. But the aim must always be to make further observations and experiments as will persuade *other* people beyond reasonable doubt. The '5% significance level' almost implies that 1 in 20 of one's colleagues is entitled to disbelieve one! What we must seek is *overwhelming* evidence that will persuade *everybody*—one doubt in 10,000 might be too many. Our first care, in Science, is to preserve the consensus from unwitting error; what is certain must be clearly delineated from what is conjectured; the continuous incorporation of merely probable results must inevitably lead to a degradation of the credibility of the whole scientific enterprise.

A branch of Science that must depend heavily on statistical inference is therefore in grave peril of error. Without the most rigorous and critical analysis of its concepts and methods it may fall to worshipping very strange gods. From the point of view of the present essay, the attempt to derive, say, psychological and sociological results by such methods appears a desperate last resort.*

It is interesting to note, however, that statistical methods arose initially in the more technological branches of Science, such as Medicine and Agriculture. In Technology, when we are bound to decide immediately between various courses of action, we cannot afford to ignore any clues obtained by statistical analysis: 20 to 1 is good odds to make a profit! For the pure scientist, the statistical

* Even physics is not immune from such errors. The story of the fundamental experiment on 'charge independence', that got into the *New York Times* and seemed to promise a Nobel Prize, but was not confirmed by a later experiment with better 'statistics', brings a blush to many a cheek!

evidence that smoking causes lung cancer challenges him to uncover a causative mechanism; for the Minister of Health it is a signal to ban the advertising of cigarettes.

Yet there may be a deep reason for the belief that scientific knowledge *ought* to be reducible, eventually, to mathematical form. Mathematics—or at least its most familiar form as algebra—is the *symbolic* representation of logical or other abstract relations. A mathematical argument is invariably to be written down in formulae, which are collections of standardised signs. The information conveyed by a formula may be limited, but it is precise. It has essentially the same meaning for any well instructed reader. Whatever may be hidden in the assumptions or definitions of the symbols, it has the appearance of being entirely unambiguous.

The expression of knowledge in mathematical dress, as 'formulae', therefore make it unexceptionally public. The need to communicate ideas from one scientist to another, and the misunderstandings that could arise in the act of translation from the internal language of one mind to that of another, exerts pressure to turn all knowledge into this form.

There is a historical element in this tendency. Modern science arose shortly after the discovery of printing. It will be argued in a later chapter that this may have been no accident—that the communications systems needed in the scientific community could not have been provided by a less powerful technique. Until recently the technology of printing was limited to the reproduction of what I have called 'formulae', i.e. concatenations of standard symbols, as in the letters of a word or in an algebraic equation. It was not possible to replicate a non-standard, arbitrary figure, such as a drawing, or a graph, except by a laborious process fraught with error. The information that *could* be made public and permanent had to be formularized for printing.

McLuhan has argued* that this had most important effects on the structure of thought and feeling. During the two or three

* H. M. McLuhan *The Gutenberg Galaxy* (1962) and *Understanding Media* (1964) (London: Routledge and Kegan Paul).

centuries of the dominance of this type of information medium, what one might call the *Laplacian Metaphysic* took command of scientific thinking. The idea that every event in the universe could be described and predicted mathematically seized the imagination of scholars. And within mathematics itself we find an extremism in favour of the 'analytical' or 'algebraic' approach, so that Lagrange even boasts, in his *Mécanique Analytique*, that he has not had to use a single diagram in his treatment of mechanics, the stars, etc.

Can we accept McLuhan's further argument in *Understanding Media* that the 'Gutenberg' phase is already on the way out? The rise of new techniques altered our basic ways of thought, and aesthetic sensibilities, which will go on changing very rapidly as further techniques are invented and applied. Will there be a corresponding revolution in scientific thought—not merely in the instruments of observation but in the very 'logic' of argument and proof?*

Consider the effects of the invention of photography and photogravure—already more than a century old. These make it possible to make information public without recourse to symbols. The artist's sketch, the schematic diagram can not only be reproduced much more easily, so that modern books of science contain many more 'figures' than their predecessors; a good photograph of, say, growth spirals on a crystal, or dune patterns on a desert, or craters on Mars, can convey knowledge, theories, ideas, almost without words of explanation. It is no longer necessary to force this information through the destructive and narrow channels of words and formulae to get it from one mind into another. In fact, the eye and brain can 'read' from a single picture of this sort what would take, perhaps, a whole book to state explicitly in symbolic or alphabetic form.

It may be objected that a photograph is merely a collection of data, which must be analysed before they become scientific knowledge in the theoretical abstract sense. But suppose, say, that we

* This is the main theme of Meredith's *Instruments of Communication*, in which so many topics concerning Science and Philosophy are discussed.

build an apparatus in which a bullet is fired into a plastic plate, and we print a photograph of the resulting impact pattern next to the photograph of a Martian crater. Will any amount of algebraic analysis tell us more than we might learn from the comparison of the photographs?

Photography is only one of the oldest of such 'media'; the possibility of television, of sound recording, data processing by computer, telemetry, analogue machines, etc. have yet to be realized—not only as techniques of scientific investigation but as the means of communication and 'publication' of knowledge. For example, physicists and engineers will sometimes make available to their colleagues the tapes of instructions for computer programs that they have devised for some particular purpose, such as for the solution of some difficult equation, or for the reduction and analysis of certain types of data. These tapes are collected in 'libraries' at computer centres, and are used directly to generate further knowledge. It seems to me that these are not just tools of research; they embody information, and play just the same role as would mathematical formulae published in books, or tables of physical data in scientific papers. A computer tape is, of course, in the form of a sequence of symbols, not unlike an algebraic formula, but this is not an essential limitation, except at the stage when it is being prepared by a human mind. It can be transformed, in the computer, to something much more complex, which is unintelligible to the bare human intellect, and can only be read, so to speak, by another computer.

What I am trying to get at here is subtle and speculative, but is implied, I feel, by the definition of Science as Public Knowledge. Our experience of the form that such knowledge can take is so limited to verbalizations and formulae—to written and printed words and symbols—that we instinctively frame our definitions of Science within the framework of such forms. Our theories and explanations have to pass through this filter, in which causes are only allowed to act singly, and effects can only be observed in one dimension at a time. A photograph, a tape-recording, an electronic device, can react to many causes simultaneously, and yet record the consequences, as a complex pattern, accurately and reproducibly. It thus permits us to entertain theories and explanations whose workings and consequences cannot be represented by symbols placed in order on the page. The very nature of scientific thought must be modified by this development. The features to which attention is drawn in this book will still remain—publication, criticism, discovery, experiment—but some familiar conventions of 'scientific method' may suffer enlargement and transformation.

* * *

Let us now turn from the notion of 'experiment' to the notion of *discovery*. This is thought to be very characteristic of science. A vulgar definition of a scientist might be that he is a man full of curiosity who turns over stone after stone on the beach, to see what he can find there. Science, they say, is *finding out* about things.

This is true, but in a restricted sense. The mere repetition of experiment and observation, the amassing of data and information, is not sufficient; scientific activity is guided by ideas, by theories, by the desire to acquire *significant* information. The observer or experimenter wishes to make a contribution to public knowledge, and therefore tries to direct his work so that it has relevance to general notions shared by the scientific world.

This relevance is easily judged in some cases, where grand theories are obviously at stake. When it was predicted that there should exist a particle with certain properties, it was obvious to many high energy physicists that they should look for this particle, and it was not very long before the omega-minus was observed. But such occasions do not always offer themselves; we do not often have the privilege of conducting a classical crucial experiment confirming—or refuting—a prime doctrine. In such a case, indeed, we do not say that a *discovery* has been made, unless the experimenter was quite unaware of the significance of his result; we refer to the Michelson-Morley *Experiment*, a brilliant technical achievement, and quite unexpected in its result, yet so bound up with theory, so much analysed and interpreted, that it is seldom called, say, the Discovery of Zero Ether Drift.

More often, the field that is being investigated is beyond the scope of reliable or well-established theories. Our definition of Science implies that there must be large regions of ignorance, where the writ of public knowledge does not run; i.e. where there is no agreed pattern of theory and observation. When we come to explore these regions, we inevitably find many new things. How are we to judge their relevance? Why do we class some of these observations as important discoveries, others as mere trivia?

The adjectives that spring to mind are 'unexpected', 'unusual', 'striking'. But these are terms of comparison; the discovery is being judged against an implied standard of what is 'expected', 'usual', 'dull', etc. How can such a comparison be made, when Science itself is silent about the phenomena that might be observed?

We see at work here an elementary psychological mechanism, the principle of mental inertia and extrapolation. Even though official Science makes no pronouncements on the subject, we may have preconceived notions of what we shall find—perhaps just the negative notion that nothing of any interest will be observed. A discovery, then, is a falsification of this vague theory. The more it breaks the monotony of uniformity, the more interesting it appears. Every physicist will agree that a high sharp peak on an otherwise smooth curve is far more exciting than a broad hump. Every biologist will agree that a specimen of what appears to be a new genus is far more interesting than a new species. An anthropologist will be far more pleased to find a tribe where inheritance passes from great uncles to great nieces than from dull old father to dull old son.

This is obviously akin to Popper's principle of falsification.* But he would apply it to sharp cases of crucial experiment, where I would suggest that one finds it much more normally as the refutation, by direct observation, of some vague and lazy generalization. Unfortunately, the haziness of the theory that the discovery falsifies gives the process very little logical power. All we learn is that life is more complicated than we had supposed—but

* K. R. Popper: *The Logic of Scientific Discovery* (London: Hutchinson, 1959).

then what? Yet it still retains the psychological power of the falsification principle; whereas a successful verification passes us by with a mere nod of the head, an unexpected discovery stops us short, rouses our curiosity, and changes our convictions by contradicting them. This change of convictions, this goading into thought, is not for us alone; it applies to everyone who learns of the discovery and has any appreciation of its significance. In other words, the 'unexpectedness' of the observation is what gives it weight as a contribution to public knowledge, and hence as a contribution to Science. Discovery is of the utmost importance in Science, because discovery, in the sense that I am here defining it, is the means by which vague, general, untested notions are made explicit and brought into consciousness for acceptance or rejection.

The above analysis of the concept of discovery is elegantly exemplified in the story of the positron, as documented by Hanson.* This elementary particle is exactly like an electron, except that it has positive electric charge, and gives rise to characteristic tracks in cloud-chamber photographs. For several years—from about 1926 to 1933—these tracks were ignored, or explained away as 'dirt', until Anderson and Blackett, more or less independently, saw their significance and gave the correct interpretation. It was then pointed out that these must be the particles that had been predicted, in an extremely subtle and difficult theoretical paper, by Dirac, the previous year.

This story shows the power of a conventional notion extrapolated beyond its genuine limits. The reason why so many good physicists failed to recognize the new particle was that they had learnt to believe that electrons were *always* negative, and the positive charges were *always* carried by the much heavier protons and nuclei. Hanson suggests some of the historical causes for the strength of the belief in these principles; yet the particular point at issue—'there are no positive electrons'—was neither logically deduced from a formal model nor empirically verifiable. There had not been any papers showing that such particles were not consistent with

* N. R Hanson: *The Concept of the Positron* (Cambridge: Cambridge University Press, 1963).

the basic laws of physics, as then formulated, nor had any search been made for them. One could not say that anybody's pet theory was falsified by the discovery–and yet is was exciting and unexpected. The objections to the idea of a positive electron were vague, almost unconscious, yet they were sufficient to inhibit the imagination of many excellent scientists. It was almost more difficult to swim against the tide of opinion than it would have been to set out to falsify an explicit theory. That is why Anderson's discovery was so important.

It is interesting, in this case, to reflect on the logical relation between Anderson's experiment and Dirac's theory, of which he was unaware. For the purpose of making a convincing case, nothing could have been more fortunate. The experimental observation was quite independent of the theoretical prediction, so that neither party could have cooked his argument or his results. It is not surprising that the identity of the experimental particle with its theoretical model was immediately acceptable to physicists, and became part of the canon of the subject.

Similar episodes have occurred in many fields of science. There are distinct phases of development, each with its own intellectual mood and climate. At first, when nothing is really known about a subject, all is vague, mystical and conjectural. General philosophical principles are invoked—continuity, abhorrence of a vacuum, atomism, the chain of being, the act of creation—based upon the extrapolation or interpolation of the conventional wisdom of other disciplines. It may be denied that any science can be made of the subject, that it is too complex or intangible for genuine experimentation and theory. The activity of the philosophers, or theologians, with their conflicting schools and abstract notions, is repellent to the scientific intellect, so that anyone who undertakes serious study may be branded a crank and treated with deliberate disdain by respectable scholars.

Then there comes a phase of discovery. Perhaps a new technical development, perhaps the skill and enterprise of a pioneering mind, gives rise to new observations, objective and repeatable. In the light of the vague general notions prevalent at the time, these discoveries are unexpected and thought-provoking, for they may unearth phenomena that seem quite inexplicable and unfamiliar. The more striking and precise these phenomena, the more challenge they present. The whole field seems full of puzzles. Instead of vague, general philosophical theories which explain everything and nothing, we find numerous contradictory and highly speculative theories, tailored to fit a few of the newly discovered phenomena but not capable of explaining them all. By now it is realized that all the extraordinary facts must have their pattern, and many young scientists are drawn into the field in the hope of winning glory there.

The next phase is called, in the jargon of our day, the epoch of the 'breakthrough'. Somebody puts forward ideas, or makes clever experiments, that indicate the general pattern of explanation. Not only does he interpret some of the phenomena; his style of argument, his theoretical models, can be adapted by others to solve numerous problems all over the field. This phase, even for those who only follow their leader, is exciting and rewarding; suddenly their science makes great progress, and becomes fashionable in the academies and in the intellectual world. But it may also be a period of conflict and some bitterness, for the new ideas meet inevitable resistance, not only from the Old Guard, governed by a conventional conservatism dating from the semi-philosophical or theological era, but also from the disappointed rival theorists who are reluctant to give up the hope of themselves providing the successful interpretation. These are the most dangerous critics, for they have more to lose than the simple conservatives.

Finally, we enter a 'classical' phase, when the remaining pieces of the jigsaw are put into their places, usually by the industrious pupils of the participants in the original revolution, drawn into the subject by the interest it has aroused among the public at large. In this phase the new theory, the new pattern of thought, hardens into an orthodoxy, which will itself be the conventional wisdom which must be broken through in the next major advance.

This sequence of events must surely be familiar to anyone who has studied the history of any field of Science. We can easily think of the Copernican and Newtonian revolutions, the work of Lavoisier, the Laws of Thermodynamics, Relativity, Quantum Theory, Evolution, the Germ Theory of Disease, Cellular Biology, etc. It applies not only to such grand topics, but to fields within fields. The subject of Superconductivity has gone through all these stages in the last forty years, with its breakthrough period some ten years ago, and is now entering a classical phase—but within that phase again we observe new discoveries, requiring new insights and new patterns of explanation, leading on to new orthodoxies.

We owe to Kuhn* the recognition and analysis of this pattern of events. We see Science thus as public knowledge at its most manifest. The climate of professional opinion at any one moment is as important as the genius of individuals in determing the intellectual history of the subject. In the early phases, this hinders progress, by its denial of scientific standing to the field, and by the unreflecting consensus of a conservative doctrine. Even in the phase of discovery, a bad theory (phlogiston, caloric—and many more technical examples) that explains some of the phenomena too glibly may be seized upon and become a public doctrine, so that the minds of independent scholars are clouded by it. It is astonishing, for example, that Rumford's famous experiment of producing heat by friction did not immediately put paid to the caloric theory, which it totally 'falsifies'; the power of received scientific opinion can be altogether too great.

On the other hand, after the breakthrough, the public excitement, the debates, the criticism, the rapid exploitation of the new ideas, all interact with one another to spread it more quickly, so that there is usually much more rapid acceptance of a convincing and successful idea in scientific circles than in the world at large. As we should say in Physics, there is a cooperative effect, and positive feed-back, so that change of opinion can occur with astonishing speed, once it has been adequately nucleated. To understand such a phenomenon, we must allow for the interaction between the intellectual and sociological modes in Science.

* T. S. Kuhn: *The Structure of Scientific Revolutions.*

Yet I do not quite agree with Kuhn in likening such events to political revolutions.* The point is that the new theory seldom displaces a true government-in-being over the whole realm. As I have indicated, the scientific consensus before the breakthrough is vague and cautious. There may be numerous speculative theories, but their relative contradictions are recognized, and they are not held in high esteem. Those working in the field may have their favourite theories, around which they plan their research, but it is not usual for one of these to become a restricting and reactionary doctrine. Experienced scientists in highly developed disciplines recognize the distortions that their own thinking may suffer if they give themselves too enthusiastically to one or another partial and imperfect theory, and learn to withhold their assent. The precursor to their revolutionary government is not an *ancien régime* but a sort of weak and anarchical feudalism.

Perhaps, as befits an American scholar, Kuhn has in mind something more like the War of Independence. There is an analogy between the phase of discovery in science and that period in the history of the American colonies when the Eastern seaboard owed a grudging allegiance to the English Crown, but the civil power did not reach far into the vast and unexplored hinterland. The Revolution not only broke the power of the Crown; it released the forces to push the Frontier westward and bring the whole Continent under control. We may think of the same process occurring in miniature, in each Territory as it moved to Statehood—the transition from clusters of settlers, independently and crudely self-governing in their localities, to a single organized central authority.†

* * *

* Perhaps his interest mainly in the 'fundamental' sciences of Physics and Chemistry may have caused him to give too much weight to paradigms that imply a total world picture. To change these was indeed as difficult as a religious conversion or political revolution.

† I recall an intellectual and scholarly Cabinet Minister using the same phrase about 'pushing back the frontiers of knowledge', at least a dozen times in an after dinner speech; it must be admitted that this cliché, for all its triteness, is an apt metaphor for the progress of Science.

To understand scientific knowledge, we must not only understand scientific ignorance; we must also understand scientific error. Science is usually reliable in its positive assertions, and good scientists know how to admit that they do not know the answer to one's question. But sometimes scientific statements are made, and accepted widely, that turn out to be quite false. This phenomenon, which is perfectly familiar to practising scientists, and to anyone who has read the history of Science, lies outside the scope of the positivist metaphysic, whose weakness is thereby exposed in all its nakedness.

In the first place, of course, we all make mistakes. In so far as scientific knowledge is the sum of all the published researches of scientists, it must contain numerous individual errors. Experiment is difficult. We are often working at the very edges of technique and observation, struggling to extract an ounce of significance and meaning from a ton of dirty, noisy, untidy goings on. A piece of research is a work of art, which may be accomplished skilfully and beautifully, or may be a throughly botched and clumsy job. Some scientists are not as clever as others, and produce careless, inaccurate results, or half-baked theories. As we shall see, there are filtering mechanisms by which the most blatant errors are kept from publication, but these could never be made perfectly efficient. The general procedure is to allow all work that is apparently valid to be published: time and further research will eventually separate the true from the false. A mistaken observation will be repeated, and the discrepancy noted and corrected; a bad piece of logic or of calculation will be reworked and put right by some other person in due course.

The scientific system, then, makes allowances for individual errors, and has its techniques for checking and correction. But these techniques are not themselves infallible. There is no litmus-paper test for truth; that is, after all, the main theme of this essay. It is more difficult for a whole group of independent individuals to be deceived than it is for any one of them, but it is certainly not impossible. In any case, life is too short for every experiment and calculation to be repeated, the discrepancies further checked and interpreted, and the record put straight. We mostly prefer to assume that everything 'in the literature' is correct, until we have good reason to doubt it, or at least until we need to rely upon it rather heavily for our own work.

But major discoveries and theories are always checked by repetition and detailed criticism: how can these sometimes go wrong? As I have suggested, we often observe some powerful ideological principle at work, some peculiar bias in the theoretical consensus. A fascinating example is the rejection of the hypothesis of Continental Drift for some fifty years after it was first put forward on a serious scientific basis by Wegener at the beginning of the century. This is a long story, which should be carefully studied by those who interest themselves in the history of Science, for there are lessons to learn from it.

As I understand it, what happened was that Wegener advanced his hypothesis on the basis of such evidence as the geographical fit between the coasts of Africa and South America, and the similarities of flora and fauna of the southern continents. This evidence is quite intelligible to the layman, and plausible enough, but it was rejected by the leading geologists of the day, who were not very interested in or knowledgeable about the geology of the Southern Hemisphere and who felt the whole thing was too radical. They were backed up by Jeffreys, a Cambridge mathematician, who calculated, on the basis of seismological evidence, that the earth must be too rigid to allow of such drastic phenomena.

The subject was, then, not quite closed, but dropped into the limbo of cranky and speculative notions. Only a few individual scientists continued the search for further evidence; in the main centres of geological teaching and research it became almost heretical to admit to belief in Wegener's hypothesis. It was not perhaps quite as dangerous as for a Shakespearian scholar to flirt with Baconism, but it was a sign of being, shall we say 'not quite sound' or 'a little lacking in critical faculty'.

Then, after the Second World War, an entirely new technique, the study of the magnetism of the rocks, began to produce quite new evidence in favour of Continental Drift. After a further series of debates, this evidence began to convince the younger generation and geologists and geophysicists, and it may now be said to be a well-established scientific theory. Some of the old school of geologists still stand out against it, but the multiplication of evidence of all sorts is now so rapid that it cannot be long before their conservatism will appear stubborn rather than cautious.

The point of this story is, however, that much of the evidence now put forward on behalf of the Continental Drift does not depend on the new geophysical techniques. Had an effort been made, 50 years ago, to look seriously for conventional geological evidence, many very strong arguments could have been made in its favour. It has not been difficult in recent years to persuade the other Cambridge mathematicians to produce calculations refuting Jeffreys and supporting the theory—and so on.

Why were such investigations not put in train long ago? The theory is obviously of prime importance, with tremendous consequences for all kinds of major geological structures. According to the picture of Science as the accumulation of the independent investigations of individuals, an imaginative hypothesis on this scale should have been the *leitmotif* of innumerable researches, which would have verified or falsified it in detail many years ago. The inhibition of these studies is a clear sign of the power of the climate of opinion, of the effective consensus of geological science having rejected the hypothesis and excluded it from accepted Public Knowledge. The influence of the mathematical argument (which was probably not fully understood by most of those who assumed it was valid) is particularly interesting; there is nothing like an abstruse mathematical proof to convince the lay public, even against apparent factual evidence.

This tale is not told simply to poke fun at the geologists; similar episodes occur in all branches of Science. They emphasize the social character of knowledge. Not only do individual scientists make mistakes in their research, or fail to see what is staring up at them under their very noses. The social and cooperative nature of Science allows of grander aberrations—just as it usually allows of more rapid and surer progress than when men work alone and without attention to each other's discoveries.

The corporate, cooperative nature of scientific argument is made very obvious by the systematic use of *references* or *citations* in scientific papers. It is almost impossible to write or get published on a scientific theme without noting explicitly all relevant preceding work by other scholars. A paper without footnotes or 'bibliography' is immediately suspect; it is an elementary sign of the crank or tyro.

Why should this be? If the logic of Science is directly from observation to theory, there ought to be many scientific papers reporting quite new experiments, or new deductions, performed uniquely by the individual research worker and brought directly to the public eye. But such cases are rare.

It is true that many references are cited out of politeness, policy or piety. It is a matter of courtesy to mention explicitly all papers by one's friends and colleagues, to give them an airing and to show that they are not to be entirely neglected. One may feel that it lends some slight import to one's own modest contribution if it can be shown to be related to a more famous and significant work. Or one may simply follow the custom of citing some early and basic paper, the foundation of the subject on which one is writing, even though it has now passed into the general currency of the science, and is to be found, in one form or another, in every graduate text. The sociological and psychological function of such references will be discussed in another place.

But even when such superfluous material has been pruned away, we are left with a number of genuine references to specific papers on which the new work is based. A scientific paper does not stand alone; it is embedded in the 'literature' of the subject. Every argument that is presented, many of the facts that are adduced, must be supported by documentation, almost like the 'precedents' of a Common Law judgement. It is not unusual to find a paper that is no more than a rehash of other people's results and ideas—a bad paper, of course, but able to masquerade as an original contribution until one strips off all the references to see what is really new. As I have already emphasized, a scientist does not merely rely upon his apparatus, his eyes and his own logical powers; to an enormous extent he relies upon other people, through their published work, through the results of their experiments, through the techniques that they have initiated and tested, through the theories that they have originated and developed. The 'bibliography' of a scientific paper is a clear and explicit recognition of this dependence.

These intellectual linkages are not entirely passive channels by which knowledge is transmitted from one generation to the next. To cite favourably an earlier paper is to give it credence, and to preserve it, so to speak, from oblivion. If one cannot change past events, and make one's intellectual grandfather unsay what he has written, one can put the record straighter, and influence the assessment that future generations will make of his work. The actual process of criticism and sifting to which all scientific papers are subjected, and by which their contents come into the public domain, will be discussed in another chapter. The system of citations from one primary paper to the next is one of the main long-term mechanisms of this process. It is another of those activities where individuals unconsciously cooperate to achieve an important corporate end.

It has become of interest to study 'trees' of citations, to see how one paper is related to another, to represent each as a nodal point, gathering together threads connecting to various earlier papers and itself originating further connections to later work. There is even an attempt to create 'Citation Indexes', listing all those papers referring back to some specific work, and to use these as a means for searching out all relevant research on some particular problem.

Whether or not this is a fruitful procedure, we learn a lot from the very idea that scientific work can be analysed in this way. The pattern of citations has its own internal logic, which is not at all the theoretical logic of the subject. The historical order and connectivity that it indicates are not at all the categories and logical relations into which the field will eventually settle. For example, some particular paper, clever but only partially correct, may dominate the subject for a number of years, and then be put right, or otherwise superseded, so that it scarcely warrants further mention except in the history books. On the other hand, a beautiful and important research may lie neglected and forgotten—Mendel's Genetics, or Carnot's Thermodynamics—and then suddenly be seen as a great well of knowledge.

Scientists, in every age, do not usually try to set their sights upon the solution of some obvious major problem and subordinate their researches to a single well-defined goal. Only the crank—or his cousin the rare genius—decides to find the explanation of Gravitation or to discover the Elixir of Life. Nor are most scientists fully aware of all that has been done on all possible aspects of their chosen field; to be too completely learned may sterilize initiative and imagination. Without necessarily succumbing entirely to fashion, they tend to work within the conventional framework of currently received notions, and to build, opportunistically and piecemeal, upon the near and familiar work of their colleagues and contemporaries. The 'trees' of citations in most disciplines are very compact, with very few linkages across the specialty boundaries. There is a gentle parochiality of much research—not the fierce and bitter defensiveness of 'schools' or 'ideologies', but a preoccupation with the problems and ideas of a relatively small group of people and an unconscious assumption that these are adequate to the work in hand.*

What is to be emphasized here is that this natural, human, social tendency has a strong influence on the intellectual pattern of Science as a whole. It is a sure way to get a laugh at the expense of the academic world to read out the titles of doctoral dissertations: titles that sound as trivial as 'The number of hairs on the moles on the ears of black-backed rats' or as portentous as 'The colleostographical chromatophoresis of palaeotonic telogenies'. The implication is that scientists study only minutiae, or that they address themselves to more and more remote and abstract notions which have no connection with reality.

But within the context of his education, and for the group to which he belongs, each of these topics has meaning and purpose

* Kuhn calls this 'normal' science.

for its author. If you were prepared to learn his jargon, and to follow the steps leading to the formulation of his problem, you would see its rationale. You would find a dozen other scientists, at home and abroad, competing and cooperating to solve the same problems and to make the same discoveries. Far from being the sum of independent, individual researches, the continuous compilation of innumerable disconnected facts, observations and theories, scientific knowledge is the joint social product of the members of these 'Invisible Colleges',* whose intercourse is through the citations that they award one another, however seldom they may meet face to face. The mechanisms of criticism, and competition, the general norms of rationality and verifiability, work as effectively within these smaller groupings as they do for Science as a whole. Whatever the ultimate significance of their achievements, they are generally just as reliable and 'true' as the grander results of major theory and discovery.

Many scientists seem to believe that the intellectual progress of Science as a whole is assured if each specialty does its work satisfactorily in this way. They believe, so to speak, in a *laissez faire* economy, not insisting on the absolute independence and personal accountability of each individual, but leaving the growth of larger corporate activities to the accidents of the market place and the laws of contract.

The danger is that the links between the specialties—the intellectual links that would bind it all together into a coherent whole—may be neglected, and let go by default. It is one thing for a new idea or discovery to nucleate an Invisible College, to exploit and develop it; it is quite another for a whole science or discipline to remain strongly polarized or 'clumped' about these nucleation centres with only weak general ideas to hold it together.

There are, of course, always a few powerful eclectic minds who will attempt the synthesis, but until they have achieved sufficient personal status they do not find a ready audience for their work.

* The apt resurrection of the name of the precursor of the Royal Society is due to D. de S. Price in *Science since Babylon* (New Haven: Yale University Press 1961).

It is much easier to join a specialty, and satisfy its cosy internal criteria (however tough these may be, in a strict professional sense) than to create interests embracing a number of these little villages of the mind.

Another danger of this strictly localized attitude is that the general pattern of development of the subject may become seriously unbalanced. If we assume that progress in an ordinary scientific field depends mainly on the number of good scientists who are active there, then it is obvious that quite important scientific problems may sometimes simply not get solved because, more or less by accident, nobody happened to take them up. The reasons why any particular scientist enters a particular specialty are exceedingly complex—in fact, almost impossible to analyse except in tautological language—but they tend to favour 'fashionable' topics. It requires a deliberate and intellectual effort to make an appraisal of a large science and to direct one's attentions towards serious problems that are not being studied by other people. The cooperative nature of Science favours a clumping effect which can only be withstood by a powerful and strongly counter-suggestible personality.

We verge upon topics that belong in other chapters—perhaps in another book. But it is clear that we cannot understand even the intellectual microstructure of Science—what it knows in detail about the world—unless we reckon with its public, social nature.

4

EDUCATION FOR SCIENCE

If [a student] does not believe the statements of his tutor—probably a clergyman of mature knowledge, recognized ability and blameless character—his suspicion is irrational, and manifests a want of the power of appreciating evidence, a want fatal to his success in that branch of science which he is supposed to be cultivating.

TODHUNTER

Emphasis upon the social character of Science forces us to ask 'Who is entitled to be heard; whose opinions must be taken into account in the consensus?' It is obvious that we cannot listen seriously to the views of all mankind on such topics as the nature of the electron, or the mechanisms of heredity. The universal right to vote, the key principle of political democracy, has no validity within the intellectual realm. How, then, do we restrict the franchise, so that only 'scientifically competent' persons are competent to pronounce on scientific matters.?

This question, stated thus so tautologically, raises one of the major formal difficulties of the whole theoretical system put forward in this book. If we are not to have a consensus of all humanity, or at least an overwhelming majority, then are we not in danger of defining the scientific community as just those who subscribe to the very consensus which we are attempting to validate? In the metaphor of Elementary-Particle Physics, it becomes a bootstrap operation, or no better than a mutual admiration society.

Put in this form, the difficulty is, indeed, insuperable; the objection is unanswerable. But it springs from the desire to achieve some absolute of validity, some criterion of 'truth', not, perhaps, of the most extreme positivism, but going beyond mere opinion and belief. As I have insisted throughout this discussion, such a criterion always seems to escape us. To make truth depend upon the counting of noddles would be as dangerous as to allow justice to depend upon the will of the majority.

Nevertheless, even when we step down from the abstract metaphysical plane, and merely insist that Science is an extraordinarily effective method for getting knowledge that eventually becomes overwhelmingly acceptable to the vast majority of those who care to study it, we still have a problem. The community of those who are competent to contribute to, or criticize, scientific knowledge must not be closed; it must be larger, and more open, than the group of those who entirely accept a current consensus or orthodoxy. It is an essential element in the health of Science, or of a science, or of the sciences, that self-confirming, mutually validating circles be unable to close. Yet it is also essential that technical scientific discussion be not smothered in a cloud of ignorant prejudices and cranky speculations.

The convention is that the scientific community consists of those persons who are able to speak its language. If you wish to pronounce upon a scientific matter, whether to propose your own theory or to oppose another man's, you must show that you are already acquainted with current knowledge in that field of study. To change the consensus, you must, paradoxically, demonstrate that you understand and accept it as it is. Generally speaking, this means that you must have suffered many years of education in the subject, and almost invariably that you have passed a series of examinations, culminating in a Ph.D.

It would not be quite correct to say that an actual Ph.D. (or the equivalent degree, according to the academic system of the country) is a necessary and sufficient condition for acceptance in the scientific community. These matters are not yet laid down quite so precisely. For example, it is quite usual for a graduate with a Bachelor's degree to go into an industrial or government laboratory, where, as a research assistant, he may acquire equivalent experience. The point is that training as a scientist, up to the publication of a first piece of research, is now, almost without exception, within existing scientific institutions, under the eyes of active research workers in the subject. The case of Einstein, who left the University with a poor first degree, and worked in the

Patent Office whilst studying and researching on his own, would now be quite abnormal.

A doctorate certifies two quite different skills. On the one hand, it is supposed to indicate an acquaintance with the elementary grammar and syntax of one's field of Science—ability to manipulate Feynman diagrams, messenger RNA, Babylonic cuneiform, or whatever it may be. In this aspect, which is our concern in the present chapter, 'graduate school' is merely an extension of undergraduate education.

On the other hand, a Ph.D. requires the successful completion of a modest but original piece of research, as witnessed by a thesis or dissertation. In this aspect, which we take up in the next chapter, the graduate student is a novice or apprentice, acquiring the conventions, norms and goals of the community which he is entering.

It is quite evident, therefore, that certification as a competent scientist is entirely in the hands of the existing Establishment. The teachers, examiners and referees are all 'authorities'—well-meaning no doubt, but incapable of treating a genuine unorthodoxy—a reversion, say, to Lamarckism, or non-relativistic physics, or deterministic electron theory—as better than a mental aberration.

This closure, this ecclesiastical tendency, of modern Science is not something to be shrugged off lightly. It does conflict with the ideal of the freely accepted consensus, and threatens the basis of its credibility. How can a system that trains its own recruits so carefully fail to fall into a self-perpetrating pattern of doctrinal orthodoxy and error?*

One safeguard has been emphasized by Polanyi. Modern Science is not nearly so fragmented into hermetic disciplines—Physics, Chemistry, Botany, etc.—as in the past. The territories merge into one another, and interdisciplinary subjects—Chemical Physics, Biochemistry, etc., are strongly encouraged. It is difficult to find any boundary lines, to sharply delineate the subject matter of an academic curriculum or research journal. The consensus or paradigm in one field cannot be closed off from criticism by well-

* Here again, the point has been made most forcibly by Kuhn in *The Structure of Scientific Revolutions* e.g. p. 163 f.

informed experts in neighbouring fields: there are always authorities straddling the boundaries. We have seen this process at work in Geology, where the influx of ideas, evidence and scholars from the new interdisciplinary field of Geophysics shook the complacent rejection of the theory of Continental Drift. This mechanism of mutual criticism and accreditation maintains scholarly standards across the whole of Science.

Nevertheless, for this process to work it is essential that the general metaphysic of rationality, empiricism and criticism should be faithfully maintained within the scientific community as a whole. In conventional language, the Scientific Method of experiment and observation ultimately provides a means of correcting possible errors due to dogmatism. But this implies that each new member of the scientific community should be conscious of the foundations of his scientific beliefs, and must acquire an awareness of the general philosophy of his vocation. Some educationalists would make the History and Philosophy of Science part of the formal education of science students. My own feeling is that these subjects have already become so academicized as to be meaningless to the average young scientist, and that these are matters only to be learnt, by imitation and experience, in his novitiate as a research worker.

* * *

Undergraduate education is not only *in* science; it is usually also *for* science. For almost all the participants, whether as teachers or as students, the process has but one aim: to prepare men and women for research work. This makes discussion of curricula and aims much easier in the Science Faculties than in the Humanities and Arts; no damn nonsense about cultural values and ends need complicate the issue!

Quite apart from the deplorable consequences of presenting to students a picture of the world from which such values are absent, this attitude is dangerous from the narrower scientific point of view. As we saw in chapter 2, the subject matter of many scientific disciplines is not entirely distinct from that of the Humanities. The explicit scientific consensus may unconsciously incorporate ideas that stem from the climate of opinion in the less 'scientific' realm. This is particularly the case in a new branch of Science, whose founders may in fact have been educated in other Faculties, whose general store of learning they may take for granted. Thus, the writings of Max Weber would be lost on a student of Sociology who lacked access to a wide knowledge of the Classics and History. To try to teach a would-be 'scientific' sociology, ignoring the older culture, would be to misunderstand and misinterpret the very consensus one was attempting to expound.

The same criterion supplies one of the main objections to highly specialized education in a single narrow subject. The general unification of all the sciences which we have seen in the past half-century has produced many sub-disciplines with claims for special recognition as distinct fields of study. But at the same time, the current consensus in any such field draws its roots from several of the older major disciplines, and cannot be taught without reference to the general principles current in several of the sciences. Molecular Biology, as a fashionable example, stems from Biochemistry, from Physical Chemistry, from Crystallography, from Genetics, from Microbiology, from Cytology and from Physiology, and requires a substantial understanding of the languages of all these disjoint subjects. To expound merely those aspects that seem relevant to present knowledge and research problems in the narrow field is again to underestimate the scope of the consensus of the subject.

The conventional ticket-of-entry to a science implies the prior existence of a substantial subject matter on which consensus has been achieved. It can only be used for a discipline in its 'normal' well-developed phase. At the beginning of any science the whole situation is much more fluid. In the era of discovery (in the sense discussed in the previous chapter) it may actually be advantageous not to have been well indoctrinated in the complacent fog of contradictory ideas that passes for received opinion upon the topic in question. A good general education—even an upbringing directed towards merely 'cultural' goals—may be enough to set a man upon a trail of thought, observation or reading leading to a new and significant point of view. He may himself have to create the whole language of discourse by which to express his ideas. One of the confusions and historical fascinations of such epochs is that there is no accepted expertise, nor are there any certificated experts, to validate one another. The new science takes shape out of the personalities of a few individuals—a professor of geometry, an Irish nobleman, a farmer of the taxes, a schoolmaster, the brother of a Duke, an apprenticed laboratory assistant, a monk, an idle young man of private fortune—who have had to educate themselves, and who recognize each other by the compelling force of their scientific contributions. As we have seen, this is a period when ideologies are rife, so that too fervent discipleship to the style and technique of a bad master may actually disqualify one in the eyes of others. It is not surprising that historians and philosophers who have concentrated upon these formative epochs have failed to take account of the collective social forces that develop with time in every major science.

Is is true that some sciences have grown out of well-established technologies, with their existing educational curricula as a consensible basis. Thus, research in Medicine and in the 'pre-clinical' sciences of Anatomy and Physiology has only been open to the qualified physician, with his training in the mysteries of his profession and its special jargon. There are still obstacles to research in fields such as Pharmacology by persons who are not medically qualified, however irrelevant the clinical applications may be to their work. The whole question of the relationship between research and practice in technological fields is, indeed, of great fascination, but must not detain us here. Let us simply note that the technical education and subsequent professional registration of a doctor or engineer gives him, automatically, a right to be heard within the circles of his science. However stereotyped and formal that education may have been, however steeped in orthodoxies and mythical pieties, it gives him something of the status of a 'man of science'. In reality, the great majority of medical men and engineers do not contribute to the consensus of their disciplines, any more than chartered accountants contribute to Economics or aldermen to

Political Theory, but through their education in the esoteric language and techniques of their profession they seem, to the outside world, to participate in its scientific development. We have already noticed the confusion in the public mind between Science and Technology; that confusion is compounded by the common education of scientists and technologists in these disciplines.

Scientific education is for research; technological education is for practice. But in the end there is no great difference between them. Old technologies are transformed by vast new accessions of scientific knowledge; old sciences develop so many practical applications that many of their graduates become, in fact, technologists. In both cases, the student must acquire a thorough understanding of the consensus of his day, whether to practise it or to change it.

In these circumstances the major task, and the corresponding problem of scientific education is easily defined; it must teach the consensus without turning it into an orthodoxy. The student must become perfectly familiar and at ease with the current state of knowledge and yet ready to overthrow it, from within. He must be carried past all the mental barriers that impeded his ancestors, and yet be ready to jump off, and push for himself when the frontier is reached.

In a deeply structured discipline such as Physics, this is extraordinarily difficult. The classical theory is so complete, so well understood, so total in its applicability that it is almost inevitably taught as a revealed system. There is not time (nor would the effort necessarily be profitable) for the student to struggle from one new idea to the next, retracing the historical path of discovery. How many false notions would he have to learn and unlearn, before being permitted to see the whole pattern as it is now understood (or thought to be understood!)? To return to a former intellectual era is impossible; the merest whisper of, say, the concept of energy, or of entropy, of light as electromagnetic waves, of a universal force of gravity, of a gas as swarm of particles, would corrupt our innocence and make us despise the achievements of our forefathers, rather than appreciate them. There is much to be learnt from a serious study of the history of Science, but it is no way to the best comprehension of Science as it exists. Kuhn* has remarked on the contrast between the science student, who reads only textbooks by second-rate scholars, and the student of the Humanities, whose curriculum is filled with original texts by the Best Authors. Nothing could bring out more vividly the consensus principle. A great scientific discovery does not exist by the moral authority or literary skill of its creator, but by its recognition and appropriation by the whole scientific community. The original papers of Maxwell, Einstein, Darwin and Mendel are no more than historical evidence for their brilliance in their day. Except for the accident, in some cases, of an elegant and witty style, worth reading for pleasure if not for primary instruction, they may be superseded without loss by the superior accounts that a century of teaching experience now provides. But the writings of a great critic or historian (not to mention those of poets and other artistic creators) are part of their own message. To paraphrase the argument would be to destroy its effect. Knowledge and insight here are private and personal, not consensible. Their words can only be read and re-read, by successive generations of students and scholars, for their direct influence upon the spirit.

Here again, in those fields that straddle the Humanistic and Scientific Faculties a middle way is essential. Where many problems are only vaguely formulated, where speculative systems vie for acceptance, where conflicting schools of interpretation or of methodology are at daggers drawn, only a catholic taste and wide reading from original sources can save the student from the worst sort of ideology and dogmatism. These dangers are, it seems, greatest in just those fields where a consensus exists over some parts, and where the protagonists of the 'scientific' approach may be tempted to give their own prejudicial account of less amenable topics.

In his discussion of scientific revolutions, Kuhn emphasized the difficulty that every scientist has of changing the paradigm—the set of basic assumptions implied by the current consensus. Surely,

* *The Structure of Scientific Revolutions*, p. 124.

one feels, scientific education should be particularly careful to avoid this dangerous rigidity. The contradictions inherent in the paradigm should be emphasized to the student, so that he may be prepared for changes when they come.

In practice, this is impossible. The perception of genuine contradictions demanding revolutionary modifications of outlook, the suggestion of possible alternative conceptual schemata—these are the most difficult imaginative steps in the revolution itself. Even to understand the nature of the anomalies to be corrected, the student must already have a thorough grasp of the current consensus. The change of intellectual climate cannot be prepared for directly, except perhaps by a rare gift of intuition on the part of a brilliant teacher. The problem is to avoid a general psychological stiffness and pride, a dogmatic and intolerant spirit, that will not permit the mature scientist to accept such changes as must inevitably occur. This ability to progress with the subject, to grow along with the consensus, is required as much in normal periods of scientific progress as in periods of revolutionary change.

But the job of the ordinary science teacher, in the first instance, is to make all plain, and plausible, to encourage the student to entrust himself freely to the basic theory. To express doubts, to utter warnings, at this stage will inhibit the confident use of the new technique, the new language. The expertise of the professional scientist is his ability to 'think physically' or 'chemically'—to transform every problem into the concepts and formulae of his discipline. Years of reading, lectures, laboratory classes and examinations are required to acquire this skill. Those years must set their mark upon the intellect and spirit of the student, making him docile to accept rather than aggressive to question what he has been told.

The importance of good teaching at this stage is obvious. The skill to make old, well-established ideas interesting and challenging; the general grasp and overview of the subject, so that its limitations can be seen and explained; the use of newly learned techniques to solve significant problems rather than trivial contrived 'examples'; the background of active research outside the teaching laboratory or lecture room—these are qualities that can save the teaching of 'fundamentals' from degenerating into an almost Talmudic orthodoxy. The very power of high theoretical science is so overwhelming; it has none of that imperfection, that bumbling, those contradictions, that arbitrariness, that even the most rigorous and logical of dogmatic theologies must present to its devotees.

The spiritual and cultural hazards of dogmatic academicism are widely recognized and allowed for in the Humanities. They exist far more strongly in the Natural Sciences—not by inculcating an orthodox view of the world, in the sense of Kuhn's paradigm, but by the picture they present of the irresistible power and automatic progress of Science itself. Here is the source of Scientism, of doctrinal mechanistic positivism, of those misunderstandings of the nature and scope of the Scientific Method which are so properly deplored by men of sensibility and sense.

* * *

For when research itself begins, all this is changed. Everything, suddenly, must be doubted and rethought. Nothing, nor anybody, is to be trusted. There is no authority but one's own intellect.

This is a most difficult psychological chasm for the student to jump. Science as taught to undergraduates tends to be logical, precise, impersonal, authoritative; at the research working face it is intuitive, uncertain, deeply felt and controversial. It is almost impossible to convey to the layman, so impressed by the power of established Science, the confusions and errors that are always with us in the process of discovery. If everything were already clear, then there would be no research to do. The power of the scientific method is not that it keeps any one of us from error but that by mutual criticism and persuasion we gradually clarify and correct each other's intuitions, until they become part of the canon of the subject. To the research man, ideas are only interesting whilst in the phase of conjecture and hypothesis, waiting to come under the clear light of reason and experiment.

Not only is there this psychological gap; there is often an intellectual break between formal undergraduate and graduate studies and research science. The little teams burrowing away along their separate tunnels seldom pause to clear the seams that lie between them. The emphasis in research is so greatly on changing or contributing to the consensus that the task of defining and expounding it often lags far behind. The scientific literature is strewn with half-finished work, more or less correct but not completed with such care and generality as to settle the matter once and for all. The tidy comprehensiveness of undergraduate Science, marshalled by the brisk pens of the latest complacent generation of textbook writers, gives way to a nondescript land, of bits and pieces and yawning gaps, vast fruitless edifices and tiny elegant masterpieces, through which the graduate student is expected to find his way with only a muddled review article as a guide. The consensus may exist chiefly in the knowledge and wisdom of experienced scholars in the field and only by implication in the published literature.

The social relations between student and professor also change. From being deliberately subordinated as an apprentice to a master whose mind he must pick and whose ideas he must make his own, the student must become independent, 'self-winding', a colleague. The two defences that he may have constructed—deferential humility or aggressive self-assertion—now become quite inappropriate in his relations with other scientists. The research student must exchange the norms of undergraduate life for those of the scientific community to which he aspires.

This important stage in the life cycle of the scholar will be discussed further in the next chapter. My present purpose is simply to indicate how it is determined by the concept of the consensus. The contrasts and contradictions at which I have hinted are deep rooted; they cannot be eliminated by precept or programme. Well-established scientific knowledge is academic, a form of orthodoxy, all the more dogmatic for its rationality; in the research phase, Science is romantic in its chaos.

It is necessary for his very understanding of what Science is and can do that the student become acquainted with it in both its moods. An education in a high pure science, ending with a first degree or after a year or two of graduate instruction, is overwhelmingly formal and rational. The picture that it paints of the world, and of Science itself, is altogether too cartesian. It tends to suggest scientistic prescriptions for all the evils of mankind, without the saving graces of humanity and error. Many students, of high moral sensibility, recoil from Science at this stage, revulsed by its impersonality and blind, crude force. Others swallow it whole, and go through life thereafter imagining that every problem could, in principle, be solved by 'the Scientific Method', just as Newton solved the problem of the motion of the heavens or Kekulé solved the problem of the structure of benzene.

A little experience of research helps correct these grandiose notions. Indeed, many young research workers become so disillusioned that they lose interest in Science altogether; it has not lived up to its dream of rigour, nor, perhaps, given them the authoritarian discipline that they ask to govern their lives. Others never do grow out of the Wellsian phase, and pursue extraordinary careers in which they proclaim the power, precision and mechanical efficiency of 'Science' in general, whilst performing marvellous feats of intuition, imagination, irresponsibility and paradox in their scholarly spheres. But most professional scientists are humble practical men who have learnt from experience the limitations of their arts, and know that they know very little. This may be another reason why the 'Philosophy of Science' means nothing to the research worker; all too often, it is a hind-sighted analysis of the classical phases of the subject, the sort of things that schoolboys are taught—living ideas turned into stone.

If, as many educationalists now think, it is important to demonstrate to Arts students 'what makes scientists tick', then the study of a formal 'exact' science such as Physics is quite inappropriate. In the early stages of learning, such a science is inevitably abstract, rigorous and 'classical'. A much more suitable topic would be, say, experimental Psychology, where many of the basic phenomena are already more or less familiar and where controversy and uncertainty are endemic from the beginning. The lesson to be learnt is not the abstract power of 'numeracy' or of logic, but that unsuspected and useful knowledge can be gained from such unpromising and complex material by imagination and curiosity, honesty and suspicion, experiment and ratiocination. As I have already remarked, the belief that all scientific knowledge may eventually be reduced to Physics and hence in the end to be subsumed to Mathematics is not well founded. To worship Physics as the paradigm of Science is to be blind to the genuine powers and methods of other forms of enquiry, and other forms of theory. This is the source of much of the suspicion and fear of 'Science' amongst non-scientists. If all Science ought to be like Physics, and if Physics is difficult, cold and esoteric, then the same attributes are wished upon Science as a whole.

Even for the student of Science and the aspiring research worker this is important. Our way of teaching the high pure sciences accentuates the positivist metaphysic by the total emphasis, in the initial years, on the well-established consensus of the past. It is our duty, if possible, to introduce more controversial topics from an early stage—for example, by a study of the applications of the techniques to less 'advanced' problems. Thus, the great debates now raging on Continental Drift, to which I have already referred, do not require a deep knowledge of Physics or Mathematics or Chemistry to be appreciated: that is the sort of topic that teaches us what Science is really like on the research frontier.

Again, we may begin to understand the importance of combining teaching and research. The conventional argument is that contact with young fresh minds keeps the research scholar alert and undogmatic; we might equally hold that the practice of research, with all its uncertainties and difficulties, keeps the teacher humble and undogmatic, and gives the student some inkling of the way in which true Science is in fact conducted.

But education and research are so intimately related that this chapter could be made to swallow the whole book. The notion of Public Knowledge implies educational institutions where that knowledge may be passed on from one generation to the next. In the revolutionary phases of his research, the active scientist creates or modifies elements of the consensus; as a teacher he recreates it for his students. The act of teaching is not merely passive; as knowledge is expounded it becomes more orderly, formalized and refined. The education of the student provides him with the basis of his consensible knowledge, which gives him common intellectual ground with his scientific peers, and allows him, in his turn, to add to the communal store. Normal Science is necessarily academic in a way that normal poetry and normal life should not and cannot be.

5

THE INDIVIDUAL SCIENTIST

Where we desire to be informed, 'tis good to contest with men above ourselves; but, to confirm and establish our opinions, 'tis best to argue with judgements below our own, that the frequent spoils and victories over their reasons may settle in ourselves an esteem and confirmed opinion of our own. SIR THOMAS BROWNE

The concept of public knowledge faces us with the following paradox: the scientist regards himself as entirely free in his research; how can he simultaneously be a cog in a social machine for producing knowledge?

This is a more special case of the familiar problem of liberty in society at large. A free citizen of a republic similarly regards himself as able to make his own decisions about his life; yet in a well-ordered society he will in fact be performing a specialist function with the same reliable certainty as an ant or bee.

The answer is, of course, that by his upbringing he is conditioned to the norms of his society; he learns to work for reward, to covet money, to obey traffic signals, to be honest, loyal, faithful, etc., so that his freedom is circumscribed by invisible glass walls within his own personality, so that he does what society expects of him without having to be told.

Similarly, the professional research worker is conditioned to the norms of the scholarly community. This does not constrain his field of study (any more than ordinary bourgeois society need tell John Doe to be a stockbroker rather than an ironmonger) nor tell him what discoveries to make in that field, but the whole style of his investigations and their subsequent publication is strongly determined by those norms. It is not enough to be intellectually acquainted with the current consensus; he must learn how to behave like a scientist; indeed, he must learn to 'think scientifically'. He must internalize the scientific attitude so that he cannot even conceive of, say, writing a scholarly paper in Zen, or recording the epoch of an eclipse by reference to the age of the reigning monarch, or pay a claque to abuse the reading of his opponent's papers.

Now it is sometimes held that the scientific attitude is so completely logical, rational and unequivocal that such self-discipline is unnecessary. One is told that the scientist succeeds by freeing his mind of cant, by being absolutely honest, by sitting down before the facts like a little child, by measuring all things, by having no preconceived notions, by observing impersonally and objectively, etc., etc. These are admirable virtues and no doubt every serious scholar, in every age, has tried to practise them. They are virtues that are strongly reinforced in some aspects and applications by a scientific career—although we all know of scientists of the highest repute whose political attitudes and moral principles are infantile, or even dishonest. But in any particular context these scientific virtues are neither self-evident nor easy to practise. Only hindsight justifies the charge of prejudice against those who did not immediately accept all of Galileo's arguments, or who were astonished and confused by Darwin's views on the Descent of Man. Honesty, lack of cant, objectivity, etc. are impossible conditions to achieve, and sometimes quite irrelevant to the business of making scientific discoveries.

What I have tried to show, in chapter 3, is that the criteria of proof in Science are public, and not private; that the allegiance of the scientist is towards the creation of a consensus. The rationale of the 'scientific attitude' is not that there is a set of angelic qualities of mind possessed by individual scientists that guarantees the validity of their every thought—as if they were, so to speak, well-tuned computing machines whose logical circuits precluded them from error—but that scientists learn to communicate with one another in such tones as to further the consensible end to which they are all striving, and eventually train themselves to construct their own internal dialogues in the same language. A private psychological censor takes over from the public policeman or parent, and conforms our behaviour to social norms. But he does not keep whispering into our ear, 'Be honest, be truthful, be objective', in a chorus of pious aspiration; he says, 'Have you checked for instrumental errors? Is that series convergent? Would anyone understand that sentence? What is the present status of that old bit of theory?' and so on.

In other words, a peculiar quality of the research scientist—and this he shares with professional academics and scholars in all the faculties—is that he has very high internal critical standards for arguments within the context of his discipline. He knows his consensus and he has to decide whether a proposed change is acceptable. When faced with the argument for such a change, whether from another person or from his own mind, he instinctively puts on his referee's mantle, and weighs it up as if he were an official representative of his intellectual community; he does not say, 'Can I believe this?' but rather, 'Would *they* be convinced by this evidence?' Far from being *im*personal he tries to be *omni*personal in his judgement.

Where does 'creativity' come in? Well, of course, a good scientist has got to have it, for it has become the definition of being a good scientist! The better word is *imagination*—the ability to construct new patterns and combinations of ideas.

But scientific imagination is strongly constrained: scientific speculation is far from idle. It must act within a well-organized framework of ideas and facts, with rigid rules of argument and proof. Even the cosmologists do not spin their marvellous webs of space and time out of mere fantasy; they use the logic of the tensor calculus and astronomical observations to construct rational systems compelling by their elegance and simplicity. It is much more like extremely academic painting or poetry, in which the art is to say something new within the official stylistic conventions, than abstract expressionism or even blank verse. The typical scientific paper is akin to a sonnet, or a fugue, or a master's game of chess, in its respect for the regulations. Many scientists delight in just this feature of the game; it is more satisfactory, and the victory is more definite, to win under such prescribed conditions. Scientific imagination is a rare gift, but its achievements are not so much at the mercy of fashion and prejudice as those of poetic and artistic genius.

Learning, imagination and critical sense—these are the three qualities which the scientific mind must possess in abundance. He must not be ignorant of the present and past consensus; he must not be blind to the possibility of changing it; he must not give credence to every passing whim or fancy. Of course, these qualities are not usually found in the same proportions, even in the very best scientists. Some men are very learned, able to bring to mind a relevant fact or notion on any subject, repositories of wisdom for their students and contemporaries. Although inhibited from imaginative steps by their attachment to existing knowledge, they are the safe moorings from which more lively craft venture forth. Some are very imaginative, having a hundred new ideas a day, of which ninety are patent nonsense, another nine are eventually found wrong and the remaining one is a winner.* By themselves, such scientists are incapable of picking out the serious words from their own jests, but they are a yeast which ferments more pedestrian minds into activity. Some scientists, alas for themselves, are saddled with such a strong critical sense that their imaginations are hobbled by the knowledge of all the weaknesses in their own arguments—but they are the anvils upon which other people's reasons are forged, and earn professional respect by the power and integrity of their minds.

Yet these qualities are not necessarily antagonistic or mutually exclusive. It requires imagination to see the flaw in conventional arguments, so as to criticise them adequately. Suspicion of facile new ideas may be the motivation for a deep search that eventually leads to a revolutionary discovery. Ignorance of the latest methods for attacking a problem may be an advantage in allowing one to look at it with a child-like eye; on the other hand, wide knowledge of many cognate fields may provide weapons for a new and 'imaginative' approach.

* Of one such man, now a very distinguished educationalist, it is said that he has slowed down to only ten ideas a day, but now gets ten per cent of them right!

But the psychological strategies that are called for in the solution of scientific problems are no more to be reduced to a formula than the strategies of life, love, war or business enterprise, and the qualities to which I have referred are not to be derived from, say, deeper inherent pre-existent factors of the mind such as I.Q., introversion, divergence, verbal facility, etc. It is true, in general, that good scientists are usually quite intelligent—in the top ten per cent of the population, say—but this only suggests that the ability to solve set-piece problems quickly is helpful in the longer process of asking and answering unsolved problems; it may indicate no more than that the right to receive an education in the current consensus of a high science is restricted to those who have shown this particular skill. The attempt to construct a psychological profile of the 'creative' scientific intellect, without reference to the training or other experience that the person has undergone in the scientific field, is ludicrous; one might as sensibly construct a profile of 'saintliness' without reference to the theology or ecclesiology of the day.

This I assert on the basis of one evident fact—that the modern scientific worker is *made*, not *born*. Just what draws young men into the learned professions and into scholarly pursuits is very hard to uncover. The factors are multifold—the prestige and fame of successful scientists, the desire to be 'useful' rather than materially successful, natural curiosity, precocious academic talent, a philosophical bent, family background, and so on. But we know, from personal enquiry and by elaborate statistics, that many scientists—even some of the best—came into the profession quite accidentally, without an early intention or a deep youthful interest in science. Although they may eventually feel that they have found their vocation, one often has the impression that many of them would be equally happy and successful as doctors or lawyers, churchmen or engineers. The world of research is manned largely by what is aptly called a Scientific Civil Service—officers of a professional bureaucracy rather than an elect of the intellect. Research is not a 'natural' activity of mankind and it took many centuries of civilization for the technique of it to be discovered;

but now almost anyone of good intelligence can be taught to do it passably, just as almost any healthy person can learn such 'unnatural' activities as skiing, or parachuting or driving a motor car.

In its heroic age, Science was created by men with the will to withdraw themselves from the commonplace notions and pursuits of their day. Many of them were actual social recluses, enabled by inherited wealth, an undemanding profession, monastic seclusion, an academic, clerical or political sinecure, to free themselves from the ordinary life around them so as to devote themselves to pure knowledge. Others, such as Gilbert, Franklin, Lavoisier, were extraordinarily capable in worldly affairs, and yet made for themselves the time to prosecute their researches. Except for a few national academies, there were no institutions offering a full-time career in science, and the general intellectual atmosphere was not particularly friendly to 'Natural Philosophy'.

Our picture of the inner scientific life is strongly coloured by the mythology of this period. We see the man of science as a lonely, dedicated personality, grappling with problems that he has set himself, sensitive to the work of others, but not primarily governed by their demands upon him. In Riesman's categorization,* he is inner-directed, and follows his own star. Whether this has anything to do with protestantism, or whether, as suggested by Feuer,† it was originally associated with liberal agnosticism, is a deep historical question. The fact remains that the scientist of the seventeenth and eighteenth centuries had to free himself from dogmatic clericalism before he could even begin work, and hence had to acquire a peculiarly strong independence of spirit.

Yet the virtues of curiosity, intellectual freedom, the questioning of all accepted doctrines, etc., which were so essential in that phase (and which are, of course, still essential to good science now), are not sufficient to make a man into a successful research worker. Those virtues are to be found in many cranky, eccentric persons, whose would-be contributions to Science are worthless

* *The Lonely Crowd* (New Haven: Yale University Press, 1950).

† *The Scientific Intellectual* (New York: Basic Books, 1963).

because they have not been subjected to the consensible discipline. In our histories of Science we celebrate the successes of that small band of warriors whom hind-sight informs us to have been on the right track; we do not bother with all the other little bands, wandering in the wilderness with strange philosophical and religious banners which would demand the same courageous virtues of their adherents.

The scientific community has always asked more of its members than adherence to some bizarre philosophical doctrine, or mere intellectual curiosity and exuberance. Through its organs of communication it brings together new discoveries and theories, and gets them criticized. In the early days, such criticism often broke into personal controversy; but eventually social and psychological mechanisms evolved by which the delicate balance between tradition and innovation, between individual pride and group achievement, could be preserved. To be in full communion within this parish, one must be ready to withstand the wounding criticism of others and to deal firmly, but without malice, with their stupidities. One must be ready to yield to an honest argument and yet not throw away a genuine point. However much public comment may have been forestalled and discounted by one's own self-critical care, there is always a risk in presenting new scientific work; every new paper is a hostage to fortune, open to the arrows of scorn or the cruel winds of neglect. There must have been a phase, perhaps in the eighteenth century, when the general scientific community was consolidated by the expulsion of these personalities who could not live within this psychological frame.*

But this process of the recruitment of scholars by voluntary self-selection from the laity was superseded in the nineteenth century. The German universities, which seem to have been the first large-scale, self-consciously professional research organizations to

* Was the reform of the Royal Society in the mid-nineteenth century, when amateur and dilettante gentlemen interested in Science ceased to be elected as Fellows, a belated recognition of this change? Professionalism here means common adherence to the norms of the consensus group, rather than pay for doing research.

offer a career of Science and Scholarship to anyone with the talents for it, depended upon a system of deliberate apprenticeship.

This system, being in fact a State Civil Service, was very formal and rigidly hierarchical. The junior posts were lowly and ill-paid. Preferment was at the mercy of the judgement of the seniors upon the products of long years of study. Recognition often came very late, and only after the publication of a thorough and complete piece of research, equivalent to the dissertation required for a higher doctorate at a British university—the sort of work one might now expect of a candidate for a Senior Lectureship, or Readership, in Britain or an Associate Professorship in the United States. Many failed these rigorous tests, and withdrew to a less demanding profession.

Here we see the scientific community moulding would-be entrants into conformity with its norms. During his long apprenticeship, the German academic internalized the conventions and criteria of the scientific life. From bitter experience he learnt not to get caught in personal controversy, nor to speculate on the basis of inadequate evidence. Under the harsh eye of his own professor he acquired the habit of checking and re-checking his observations, of writing accurately and impersonally, and of being the foremost critic of his own work. The carrot of a juicy Chair drew forth his utmost of energy and imagination—yet always within the constraints of the discipline of his seniors.

In its day, the German academic system was the admiration of the world, and extraordinarily successful as a medium in which scientific research flourished. I do not say that a large proportion of its members were not, by nature, highly gifted and dedicated men; but also by their nurture and employment they were encouraged to contribute much more purposefully to the scientific consensus than the more dilettante English and French.* With Alexander von Humboldt as their Moltke, they constituted almost an officer corps trained to collaborate and compete in the battle against ignorance.

* See J. Ben-David, 'Scientific Productivity and Academic Organization in Nineteenth Century Medicine': reported in *The Sociology of Science* (eds. Barber and Hirsch) pp. 305–28.

There was a price to pay. The individual scientist was no longer working solely for his own amusement, nor for the abstract advancement of learning, nor for posthumous glory; he was also seeking personal promotion. As we are now fully aware, this is the catalyst for the release of vast floods of tedious, prolix minutiae, impressive for quantity if seldom for quality; 'publish or perish' is not a new cry. And the tone of academic life was often repressive, spiteful and pompous, the ultimate rewards, in status and salary, being large enough, but too slow to come.*

The mechanisms of character formation which drove the German academic system were paternalistic—often patriarchal. The power of the *Professor Ordinarius* over his students and assistants was almost that of a Roman father over his family. Upon his recommendation depended all hope of promotion. The only escape from his despotism, however benevolent, was by being elected to a Chair, to become a despot oneself.

Here was the social situation within which the Freudian theory was evolved, and to which it would seem to apply with peculiar force. The student seeks to supplant his intellectual parent—but cannot fail both to admire and imitate him. The 'scientific attitude' was thus passed from generation to generation of scholars, as a very firm and self-conscious mental posture.

It has always been easy, specially for gentlemanly British dons, to sneer at the industrious pettifogging of German scholarship at its worst. But this is only the unimaginative extreme of its major virtue of critical rigour. The German system emphasized the social character of *Wissenschaft* (which included Historical and Classical studies as well as the Natural Sciences) by making sure that every new discovery or theory was thoroughly examined and exhaustively tested before admission into the consensus. The virtues that it most strongly encouraged were those of painstaking care, loving attention to detail, precision of language and of argu-

* For evidence on this, let me recommend Ernest Jones' *Life of Sigmund Freud* (London: Hogarth Press, 1953–7) and a moving essay by Max Weber, 'Science as a Vocation', reprinted in *From Max Weber* (eds. Gerth and Mills, New York 1958) pp. 129–56)—two genuises who suffered the pains of the system to the full.

ment. No side turning was to be left unexplored, no gaps were to be tolerated in the logic. The great era of German Science (which died in 1930 and has not been revived) was not only an age of experimental, observational and textual discoveries; it was the age when the foundations of Pure Mathematics were drilled deeper, to reach the firmer bedrock of formal logic; it was an age of the treatise and *Handbuch*, and of scientific and technological education.

It is the argument of this essay that such attention to observational accuracy, logical rigour and encyclopaedic detail is quite as essential to Science as imagination and inspiration. Without these 'Germanic' virtues, Science would disintegrate into schools and sects, prophets and their coteries of disciples.

But the enforcement of high critical standards by distant, anonymous authorities and institutions, such as those of editors, referees and review writers, is psychologically impracticable. The public peace is not preserved by the abstract power of the impersonal judge and vigilant policemen; it depends in detail upon the conscience of the individual citizen, moulded in childhood by the direct and personal influence of stern but loving parents. High critical standards must become part of the intellectual conscience of the scholar; he must acquire the psychological strength to withstand the temptation of shoddy thinking and tawdry brilliance. The paternalistic upbringing of the German academic of the old school gave him this moral stiffening in its most puritan mood.

He also acquired from his intellectual parent/master a vision of the Philosophy of Nature to which his work was to contribute. For the born scientist, this is the inspiration of his life, transforming it from a career into a vocation.* In the German system this ideal was transferred from one generation of scholars to the next, from the mature professor to his immature students, by precept and by example. Their professionalism was thus always imbued with a vocational spirit, which gave it significance and inner meaning.

* For this vision at its most humble and pure, let me recommend some of the writings of Faraday, whose career is also of great interest as arising out of an old-fashioned apprenticeship as laboratory assistant to Davy, whom he eventually succeeded at the Royal Institution.

Modern Science and the modern scientist were invented in nineteenth-century Germany; nowadays, of course, every nation on earth is striving to cultivate these exotic crops. It is interesting to observe the variations of technique that have entered as the 'Ph.D. system' has spread from one culture to another.

In Britain, for example, the aristocratic tradition of Oxford and Cambridge gave the young don a position of financial independence quite early in life. Formal education ceased after the first degree, and it was a point of honour for the supervisor of a research student not to interfere with the self-improvement and maturation of the young scholar. He was expected to choose his own problem, set about research on his own and quickly become a scientist in his own right. In other words, there still persisted the ethos of the phase in which the scientist was a self-appointed and dedicated amateur rather than a trained professional; until recently it was not thought too badly of an academic if he did no research at all—for who could blame him for failing to be called to such a vocation?

In the British environment the born scientist could, therefore, develop easily, without being conditioned or constrained by jealous seniors or by a strong intellectual orthodoxy. Critical standards were maintained by competitive scholarship of Fellowship examinations in the student phase, by considerable competition for good university posts and in the top ranks by the reward of the honorific title of Fellow of the Royal Society. But the middle ranks of British Science lack the sheer professionalism of the German scholarly corps. They feel impelled to imitate the casual, lackadaisical style of their abler contemporaries, without the brilliance to carry it off. They do not appreciate the value of thorough, painstaking, if pedestrian investigations, the observation of details, the collection of facts, the compilation of data, as bases for the work of more inspired scholars. Despite the old-boy networks that web British academics socially and institutionally, they do not collaborate freely, and there is little feeling of Science itself being a collective enterprise in which one should play an allotted role.

Other countries, such as Japan, copied the German system faithfully, but lacked an existing tradition of high Science to provide the necessary leaders. Several generations of industrious but uninspired professors succeeded one another, producing vast quantities of tedious and irrelevant papers, before a few brilliant men, supported by their international reputation, could rise to the top and set higher standards of education and critical evaluation.

By contrast, in Germany itself the Nazi abomination murdered and exiled the leading scholars, repudiated logic and liberality and destroyed the soul of their academic system. When the young men came back from the war, they sought to rebuild it. But the famous Chairs were empty. No amount of solemn lecturing, reading of books, experimenting and writing could replace their wisdom and experience. There was no old man dozing in the front seat of the seminar room, waiting with an innocent-seeming question to prick a bubble of conceit, or with a word of encouragement to set fire to a modestly concealed imagination. For the past twenty years, German Pure Science has been bulky but flabby; it has lacked the inner tension between imagination and criticism, between the speculative and the factual, which was its previous glory. I know of no better demonstration of the importance of the social element in scholarship. The persons and institutions of high Science cannot be created out of Baconian principles and a supply of apparatus; they are not like factories. A mature scientist takes decades of training, and is the heir of subtle intellectual traditions. Academic institutions are governed largely by unrecorded principles, handed on from father to son, from master to pupils, in the intimacies of the seminar room, the study and the laboratory.

The introduction of the German system into the United States may be dated from the foundation of Johns Hopkins University—'a Göttingen in Baltimore'—in 1876, although a number of American scholars had previously studied at German Universities. But something happened, in the more liberal and permissive social climate of the New World, to change the psychological character of the scientists that were created.

For example, graduate education in the United States has become very deliberate and organized, with a formidable load of courses to be attended and examinations to be passed before research is allowed to proceed. To some extent this somewhat rigid and rigorous system grew up to remedy the deficiencies of undergraduate education; but it has become an institution in its own right, quite different from the haphazard attendance at lectures and seminars expected of the young German scholar.

An American graduate school gives a professional polish to the language and techniques of research. By taking formal training up to the very edge of the unknown, by teaching the most up-to-date methods and the latest discoveries, the young scientist is acquainted with the current consensus almost before it has been achieved.

This has it values—but also its dangers. The American scholar, however clever or stupid he may be at heart, is not ignorant of his subject; he does not go pottering on with antiquated methods and ideas, scorning powerful new tools, saying (as I have heard it in the Cavendish) that because Rutherford could get along without quantum mechanics, so would he. But it also lends itself dreadfully to the sway of fashion. A new idea, not fully worked out or universally accepted as public knowledge, is often taught a little too dogmatically by enthusiastic champions to not very critical students. The cult of progress, of the latest thing, is strong in America, and the natural swing of opinion, interest and undiscriminating acceptance is reinforced by the too rapid incorporation of half-baked and speculative notions into the canon of the graduate school curriculum. If, in the end, the only true education is self-education, there may be much unlearning to do in later years. An orthodoxy enshrined in books is less compelling than the psychological tyranny of the lecture course, whose inner contradictions may be shuffled over and hidden by the inevitable (and glorious) ambiguities of the spoken word.

On the other hand, thesis work for the Ph.D., although taken quite seriously, is not perhaps so tough and critical as in the old German system, nor even as in Britain. This is a dangerous topic on which to generalize, as the standard is so variable from subject to subject and from school to school. It is noteworthy that the practice of having an external examiner for the dissertation—i.e., an expert from another university to report on its scientific merit—which is universal and compulsory in Britain, seems to be unknown in the United States. The standard of a Ph.D. is therefore the standard of the University that awards it, not some general standard recognized by the whole scientific community. It is not surprising that many American Ph.D. theses do not really attain the level of being publishable as scientific papers in reputable journals and that their expository style is often so poor.

It seems to me that many American Ph.D.s, in spite of the rigour of their formal training, do not acquire a very critical attitude towards their own and other people's work. Although he may seem to be using the very latest and most powerful methods, and has learned to avoid obvious mistakes of technique, algebraic errors, etc., the student is not conditioned to watch out for the logical hiatus, the falsely excluded middle, the verbal ambiguity, the divergent mathematical formula, the alternative explanation. Turgidity and verbosity stuffed with technical jargon may mask the problem to be solved, and the nature of the proposed solution may be muffled under a blanket of half-meaningful, solemn and portentous sentences.

I do not mean at all to play the chauvinistic game of sneering at such feeble stuff. Scientific research is an intensely demanding art, which comes easily to very few men, and the best American science is of incomparable virtuosity. But the American style of graduate school has come to replace the German *Institut* as the model system for the training of young scientists, the world over, and we must be aware of its defects as well as its virtues.

The major weakness is the attempt to teach this delicate and subtle activity by mass methods. It is all too easy to enlarge the classroom, add a few more benches and lecture to twice as many graduate students as one can know by name. The allotment of grades by tests and written exercises may appear to be an adequate substitute for argument, discussion and personal assessment. A clear lecture course on the latest developments may seem as valuable as the labour of disabusing each student, in private, of his misconceptions and misunderstandings of the literature.

But scholarship, for all its social aspects, is an intensely individual activity. The young scholar *must* learn to work on his own. The graduate school often has an atmosphere of the broiler house, of forced feeding. The course work is too rich a diet, and the knowledge it contains has been jelled too soon. The method of examination tends to make the student credulous, and distrustful of his own powers of comprehension of all these deep mysteries; he does not learn early enough that Homer can nod. He does not acquire that deep suspicion, counter-suggestibility and independence of mind that are so essential in scientific work. It is not so bad, perhaps, that he should be 'other directed'—which all social beings must be—but that his ideas of what will please his contemporaries and professors are success in the tests and exercises, ability to play the fashionable game, facility with the jargon, to be followed somehow by a 'breakthrough' that will bring fame. He does not see himself as engaged in a larger struggle with ignorance and error, as a member of a great movement, as a contributor to man's understanding of nature. The graduate school, by its mechanization of learning, has thrown its philosophy out of the window.

Of course, the research training of the thesis does something to redress this. But where this is done as part of a larger team, as is now very common in certain fields, the scientific spirit may not be very apparent. It is very important for the student to carry out a nearly independent investigation of his own, starting from the idea that a problem might exist, through its formulation and solution, to the final publication of the results. To be a member of a team directed by a distant and very busy leader, building just one technical link in a complicated experiment, is an inadequate apprenticeship to the art; it is as if the pupils of Rubens were to be accounted artists after five years of painting-in the buttons on his larger compositions. The fact that the apparatus is ultimately designed to catch a few omega-minus particles, or some photons from a Magellanic cloud, does not make the job of wiring its logical circuits more 'scientific' and less 'technological'. High technical standards may be achieved by the student, without a grasp of the deeper intellectual issues.

The strength of the old German system was the way in which the spirit of enquiry was passed on as an oral tradition. The enormous expansion of graduate studies and scientific research the world over since the Second World War has been too rapid for this spirit to percolate from the few centres of excellence to all the institutions now engaged in the training of scientists. The prejudice against the creation of personal 'schools' around distinguished scholars does not favour the handing on of high critical standards where these exist. It sometimes seems of Ph.D.s of mediocre universities that they have no concept of Science as a collective enterprise, and no criteria by which to judge their own efforts except those of fashion and novelty.

Does this matter? Will not Science grow just the same, as brick is piled on brick? It is true that further experience will mature them. The custom of spending several years as a postdoctoral fellow in a leading research school gives many a young scientist the opportunity to come into contact with higher standards, both of imagination and of criticism.

Science may err, but it is, after all, self-correcting. The appeal to experiment and logic is not vain. Even though we cannot positively assert that there are 'objective' criteria to be satisfied eventually by our theories, all our experience over the past 300 years makes this a reasonable belief upon which to base our studies. Scientists may make mistakes but truth must surely triumph. The pretentious, the feeble, the baseless speculation, the trivial irrelevance—all these will eventually be forgotten and buried in the library stackrooms. The process of critical evaluation and the principle of the consensus are powerful enough to deal, as they have dealt in the past, with all such follies.

There are, moreover, centres of such acknowledged excellence, in all the major advanced nations, that the gradual diffusion of genuine scholarly standards is continually occurring. The loss to German Science in the 1930s was a tremendous gain to Britain and America. Even if the exiled scholars had in fact done no more research, they were able to convey to their new pupils and colleagues the great tradition to which I have continually referred.

The restless journeyings of scientists, their conferences, sabbatical leaves, summer schools, brain drains, etc. serve to unify the scientific world, and give glimpses of the best to the most humble toilers in the vineyard. The criminality of cultural barriers such as the Iron Curtain is that they prevent these contacts and thereby hinder the achievement of the consensus of public knowledge to which we are all devoted.

The manifest internationalism of Science is not a bourgeois or communist conspiracy; it is not mere sentimentality about the Brotherhood of Man; it is inherent in the very nature of Science itself, which must always seek to encompass the largest public for the knowledge it aspires to. Having given our splendid discovery to the world, we find it intolerable to think that there might be someone, in Milwaukee or Sverdlovsk, in Pernambuco or Chittagong, who ought to know about it and is prevented by purely political obstacles from reading about it or giving it adequate criticism. This is something that non-scientists do not understand—that anyone who works in the same scientific field, who can use the same technical language, who has faced the same problems, is a colleague and comrade. It is partly a matter of having to struggle with a common enemy—Nature—but I would ground it upon the consensus principle itself. Internationalism is a primary principle of Science, demanded by the inmost law of its being.

To appreciate this, one has only to visit a scientific laboratory in a politically or geographically isolated country. It is not that the library lacks the proper books, or that the journals arrive a little late; it is the absence of contact with the current *informal* consensus, of conversation with genuine colleagues or visitors with wild new ideas, of a reliable assessment of the quality of one's work. Scientific work is only meaningful in the social context of the scientific community.

* * *

In the above pages I have gone to some lengths in criticizing some contemporary features of the scientific life. Perhaps, in a philosophical essay, one should be cooler and less engaged. But philosophical principles are supposed to be guides to action. One of the virtues of the consensus definition of Science is that it helps us to see what we are training scientists *for*. By linking the intellectual, the individual and the communal aspects of research and scholarship, it provides us with something of a moral or ethical philosophy, as well as the rational metaphysic discussed in previous chapters. It is very important to be able to give better grounds for criticizing the products of some particular academic system or graduate school than purely technical defects in their education—but this is only one example of the sort of value judgement or ethical analysis that we now have at our command.

Moreover, we must be continually concerned about the quality of the training of young scientists. Civilized societies are nowadays committed to spending something like one per cent of their resources on research, which could mean that something like one in a thousand of the population must become a professional scientist. Most of these men and women, however admirable, sincere and conscientious they may be, are in Science as a career rather than a vocation. Unlike the gentlemanly dilettante of the past, they cannot withdraw from it if the call does not come. Like the clergy, they may often be trapped between simple material desires and the demands of the religious life. It is not merely a matter of making their scientific work more skilful; we must also make their lives more meaningful, and preserve the scholarly system itself from the effects of disillusioned careerism, spiteful jealousy, hypocritical time-serving, corruption, unprincipled ambition and other evils that may arise. The whole argument of this book is to emphasize the social character of the scientific life. The health of the scientific community depends upon the details of the social transactions of its members and their acceptance of the principles upon which the conventions of the community are based.

Some of those conventions are easily learnt. The general principle of giving recognition to priority is deeply respected in the scientific world for obvious reasons. Ideas and discoveries are the only products of our labour; an idea that is already public, or learnt from someone else, has cost us no labour, and is therefore valueless:

insistence on priority is necessary to prevent plagiarism and fraud; it is the signature on the title deeds of our achievements. It is precisely the public nature of scientific knowledge, its freedom, its communism, its lack of copyright and patents and other restraints upon its use, that makes this so important. The alchemist kept the secret of transmutation, to make a private hoard of gold; the scientist, in a sense, publishes the secret in return for a million pennies of recognition from those who use his technique.

As Merton* has pointed out, there is no real mystery about the bitter conflicts over priority that sometimes arise; the recognition of originality is at stake, and ordinary personal pride can inflame a fancied wrong into a sordid dispute. But why should *originality*, in the sense of independent discovery, be the only claim to recognition in Science? What is the basis of the convention that a scientist is to be judged upon his published work in all matters involving jobs, prizes, promotion, honorific titles and other rewards?

The answer is simply that recognition is given for—being a good scientist. One's *job* is to produce original published work, and hence to contribute to Public Knowledge. Concede that the consensus principle is at the heart of science and the problem scarcely exists. Unpublished work cannot be assessed, and does not satisfy the condition of being public. Unoriginal work is otiose, and speaks for no skill beyond that of an amanuensis.† It is true that there are, as I shall argue in the next chapter, other ways of contributing to Public Knowledge than by primary papers. For example, the writing of reviews and textbooks is scarcely taken into account in the assessment of scientific reputations. But this relative lack of emphasis on subsidiary aspects of scientific activity does not conflict with the general principle.

* R. K. Merton 'Priorities in Scientific Discovery: A chapter in the Sociology of Science' reprinted in *The Sociology of Science* (eds. Barber and Hirsch) pp. 447–85.

† I cannot fathom a remark by Hagstrom (*The Scientific Community*, p. 12) asking why scientists should not simply amuse themselves by solving problems that have already been solved. The analogy with mountain climbing is false: it is impossible to solve a problem, in the fullest sense, whose solution is known. Every student of mathematics knows the difference between a real problem and an 'exercise'.

On the other hand, as emphasized by Storer,* merely successful 'role-performance' in the scientific community—by teaching, sitting on committees, being 'useful' administratively or politically —does not gain much recognition. What this proves, I think, is that the scientific community is not nearly so troubled about its own well-being as some of the sociologists of Science seem to think. It is not a tribe, or an industrial corporation, trying to maintain its own stability and continuance as a social entity; it is a voluntary association of individuals dedicated to a transcendental aim—the advancement of knowledge. It is precisely the continued preservation of the principle of rewarding only 'original contributions' that indicates the vitality of that aim. The sort of anthropological analysis of the social system of Science attempted by Hagstrom and Storer, in terms of an exchange of 'contributions' for 'recognition' makes no sense without the explicit acknowledgement of the perfectly clear ideology and metaphysic that scientists, consciously or by unconscious tradition, are in fact obeying.

Some of the other norms or conventions of the scientific community, as set out in the writings of Merton, Barber, Storer and Hagstrom, may also be understood without great difficulty as consequences of the consensus ideal.

Thus, the norms of *Universalism*, *Organized Scepticism* and *Communality* scarcely need further elucidation. But consider, for example, *Humility* and *Disinterestedness*; why should scientists be expected to take neither pride nor personal profit from their achievement? The answer, as I see it, is that a contribution to Public Knowledge has to be persuasive. If it is put forward in a bombastic 'puffing' style, or if there is a suggestion that the author is being paid in cash for his opinions, it loses an immediate claim to credibility. As I have already remarked, the abstract, impersonal style of conventional scientific communications is an attempt by the author to make his work already seem part of the consensus. Lack of humility, or an apparent ulterior motive, would be signs that the work could not, as we say, 'speak for itself', and that we

* *The Social System of Science*, p. 26.

should watch particularly for evidence of weakness or even fraud in the argument. An author betraying these symptoms is treated not as morally delinquent but as having failed to internalize the critical intellectual standards of his Invisible College, and hence as suspect in his scientific judgement.* Because so much Science has to be taken on trust, one must be particularly scrupulous in one's writings—whatever sort of a scoundrel one may be in daily life!

But moral principles, norms and social conventions only become interesting when they begin to contradict one another. The objections to secrecy, in the light of the consensus principle, need not be rehearsed; we all understand the way in which this conflicts with the norms of national security, corporate profits, etc. Not to publish what ought to belong to the consensus is a crime against Science as such, and can only be justified by the demands of a social system with other ends.

In practice, however, subtler problems arise in the course of ordinary scientific life. One of the decisions that must be taken by every research worker is when to stop his investigations and write up and publish his discoveries. Some men are perfectionists; they never feel quite confident that they have tied up all loose ends and entirely proved their theories. Others (much more common) rush into print with the notebooks of the previous day's experiment, assuming, hopefully, that everybody will want to know about their astonishing discoveries.

There can be no firm convention governing this sort of decision, but the general principles involved are easily stated. In the first place, research is useless unless it is eventually published. There can never be an end to all the implications of a particular discovery, nor can every objection to a theory be entirely countered. One should, therefore, publish the work when it has reached such maturity as to be reasonably self-contained and self-consistent, and not open to definite objections. Let other people have a go at it then.

* A distinguished geophysicist once remarked in jest that nobody believed Wegener because he obviously had a bee in his bonnet about Continental Drift, whereas Jeffreys, a shy and retiring man, obviously had nothing to gain or lose by it!

On the other hand, premature publication is one of the curses of modern Science. Many scientists are so obsessed with the fear of being 'scooped' or are so anxious to notch up a good score of publications that they issue a long succession of scrappy communications instead of waiting until the work is complete and clear and can be written up as a whole. From the consensus standpoint, this is wasteful and tiresome. Each successive partial communication is demanding to be criticized and accepted but does not carry its own full justification with it. It is an abuse of privilege to claim the ear of the profession for what is, as yet, 'interesting', thought-provoking, speculative, but by no means fully established. At the moment that he is doing a particular piece of research a scientist is, so to speak, the champion of mankind against Nature on that field, and it is his duty to carry each round through to some sort of decision. It is an impertinence to keep shouting one's success after each thrust and parry, or to suppose that other people really care. The consensus itself does not progress by infinitesimal steps. To carry conviction, to overcome mental inertia and prejudice, one does much better to wait until one has built up a strong case for a substantial advance—a camel to swallow rather than a gnat to strain at.

Consider again the problem of deciding what research to do. If one knew what questions *could* be answered, one would not need to ask them. Many scientists, in fact, merely follow the line of least resistance, continuing to pursue the same sort of research, in the same field of science, throughout their lives. When this culminates in the elucidation of the structure of haemoglobin it is labelled heroic persistence and rewarded with a Nobel Prize; very often it is not unreasonably dubbed pedestrian and plodding.

But at any given moment there are always a number of scientists looking for quite new problems, and nearly new fields of research—whether because old trails have apparently petered out or, because newly graduated, they need to set themselves up in a subject. It is natural that they should seek entry into the field where the returns seem most promising—and that many should take the same decision at the same time. As we have already remarked, there is a 'clumping' effect, tending to favour a small number of 'fashionable' subjects rather than spreading the effort over the field as a whole.

Hagstrom* has discussed the phenomenon of fashion in Science at some length, and has elucidated many of the social factors by which prestige is allotted to various fields. Yet we must be careful that the clumping effect should not be confused with such purely imitative social behaviour as fashion in clothes. The simultaneous decision of many scientists that a new field is 'promising' may be fully justified; the defect, from the point of view of the orderly advance of knowledge, is that there is no means of preventing too many of them from crowding into it all at once. In the language of Control Engineering, there is an 'overshoot', and the stabilizing counterforce—the bitter competition that develops in the over-subscribed field—is slow to act.

The fact is that the current consensus carries with it not only knowledge of problems that *have* been solved, but also strong hints and opinions as to what problems now *could* be solved. A major discovery, of theory or of technique, implies the possibility of rapid progress in the solution of many old difficulties. The individual scientist, seeking not meretricious prestige or rapid promotion, but the furtherance of his subject, rightly judges that his powers may be best expended in this work of exploitation, rather than in some backwater that seems to lead nowhere. Or he may believe that by mastering the new technique he will be able to make progress with his own old research. It must always be remembered that scientific research is a very difficult art, in which ninety-nine per cent of frustration and perspiration is not always balanced by one per cent of inspiration and elation. It requires peculiar self-confidence and will-power not to follow a line that promises relatively easy returns.

This is not to deny that the judgement of the promise of a field is often shallow and faulty. Some people somehow think that by retracing the steps of a famous discovery, or repeating it with slight variations, they may make comparable contributions. Others suppose that anything done by a big name is likely to be worth imitating. Others, again, have never learnt that all research is in-

* *The Scientific Community*, p. 177.

teresting—and all research is also dull—regardless of the field, and imagine that they will especially enjoy the atmosphere of excitement and glamour that they have heard of (mostly in newspaper articles!) in a new field.

What I would emphasize is that this sort of reasoning, for all its defects, is not mere rationalization of the desire to be in a field of high status. Such feelings are present; it may be that they are becoming more and more dominant in the world of Science; but the ethical imperatives of the consensus principle are also active, and should not be prematurely explained away.

An active research scientist is continually facing similar problems. Should he collaborate with a senior man? How far should a controversy be carried? How much direct guidance should he give to a research student? Would an 'acknowledgement' of help received be adequate, or should he invite his assistant to be joint author of a paper? Should he take time off from experimentation to write a review article? Has he sufficient ideas to justify the expenditure of large sums of somebody else's money on equipment? Should he turn aside from 'fundamental' research to exploit a practical application?

In everyday life we learn to solve such conflicts by appeal to experience and to principle. In childhood we profit particularly from the instruction of our parents, so that in many cases we only know that it is 'right' to do as we should, be honest, truthful, considerate, etc. without detailed reflection on the principles at stake. The ethical principles of the scientific life are also learnt from our seniors—our professors, research directors and colleagues.

The defect of an oral tradition is that it can change imperceptibly from generation to generation, and cannot easily be transplanted from one culture to another. *Quis custodiet ipsos custodes?*: the apprenticeship system fails if the masters themselves do not understand their craft. To prevent a drift into decadence, or corruption, it is essential to anchor conventional behaviour upon articulate rational principles.

I do not believe that such decadence or corruption is rife in modern Science. There are, as I have suggested, many unconscious self-correcting forces preserving the integrity of scholars against the spiritual temptations in becoming a 'Fifth Estate' of the realm. Yet the history of the Church offers many examples of the extraordinary way in which explicit principles of virtue may be corrupted and totally perverted. It does no harm to give a tug now and then at the cables, to see whether our moorings are holding.

It is disquieting, therefore, that the general problems raised in this chapter are given so little serious discussion. The books and articles on the 'Science of Science' seem to concentrate mainly on who ought to get how much out of which pork barrel, or whether a scientist can be happy as an administrator (is the answer, 'Yes, a very bad scientist in a very good administration', or is it 'a very good scientist in a very bad administration'?), or how strong your neuroses must be to give you a bright idea and such like eminently practical topics.

A start has been made in the writings of the new school of sociologists of Science, to which I have referred a number of times in this chapter, especially Hagstrom's excellent book, which is entirely authentic in its depiction of the opinions and behaviour of top scientists in the United States, and suggests most interesting sociological interpretation of this behaviour. But it takes certain norms of the scientific life as arbitrarily prescribed, and shows the individuals as apparently governed only by a wary and calculated self-interest within the framework of these norms—rather like children enjoying themselves as best they can within the inexplicable rules laid down by tiresome adults.

There is fine talk nowadays of teaching Science to art students, and showing them 'what makes scientists tick'. My real fear is that we do not teach what Science is to *science* students—that we have largely lost the feeling for a Philosophy of Nature and do not understand the rationale of the procedures by which that Philosophy may be established and enlarged. The principle of the intellectual consensus provides an explanation of many of these procedures and a justification for critical standards and behavioural norms that might otherwise seem arbitrary and outmoded.

6

COMMUNITY AND COMMUNICATIONS

Praise is the tribute which every man is expected to pay for the grant of perusing a manuscript. SAMUEL JOHNSON

A scientific laboratory without a library is like a decorticated cat: the motor activities continue to function, but lack coordination of memory and purpose. Ultimately, all our elaborate apparatus and skilled technicians exist only to add a few more pages to the books on the shelves.

A scholar is a man of the pen; the making of books is his vocation. Many scholarly disciplines seem to be but the making of books out of books—and yet more books from these. The library is then the quarry of his materials, an aggregation of documents, letters, manuscripts, poems, folios, editions, prints, newspapers, essays, sermons, reports, treaties, charters, editions, fables, etc., etc., to be mined, sifted, sorted and reordered. Although his prime interest is in men and their affairs, the humanist must learn to love books and libraries as the geologist loves rocks and mountains, or the astronomer loves his telescopes and observatories.

But the feeling of the natural scientist towards books and writing is ambivalent. Since Bacon's denunciation of mere book-learning, he has been taught to return always to observation and experiment, to distrust what others write and to believe only the evidence of his own senses. He himself writes only to stake claims and register formally his own discoveries and he reads only to catch hints of other people's ideas that may be incorporated in his own work. 'Writing up' his experiments seems a tiresome and boring task, often done out of laboratory working hours, in the evenings or on study leave; 'catching up with the literature' seems an unending labour whose product may be no more than a dusty pile of unread journals and reprints on the office desk.

A major consequence of my present thesis about the nature of Science is that the 'literature' of a subject is quite as important as the research work that it embodies. An investigation is by no means completed when the last pointer reading has been noted down, the last computation printed out and agreement between theory and experiment confirmed to the umpteenth decimal place. The form in which it is presented to the scientific community, the 'paper' in which it is first reported, the subsequent criticisms and citations from other authors and the eventual place that it occupies in the minds of a subsequent generation—these are all quite as much part of its life as the germ of the idea from which it originated or the carefully designed apparatus in which the hypothesis was tested and found to be good. To describe scientific research work only up to the moment when each paper is published is like attempting to describe a human community by depicting the life of each individual up to the age of puberty, without reference to those years of maturity and responsibility that follow. The progenitor of a scientific paper is like a parent, whose early influence is decisive in the character of the child, but who cannot determine the career of his offspring in the adult world.

* * *

A scientific library is not primarily a quarry, nor a factory, but a store. It is the 'memory', in which each item is continually being rewritten as new results are transferred to it. Although the stacks of volumes in a large scientific library may indeed provide material for the researches of the historian or chronicler of a subject, this is not their primary function. When a scientist consults an article in a back number of a scientific journal he is not seeking to know what the author happened to be thinking about at that particular epoch; he is looking for evidence as to what he himself should be thinking now on that subject. He is making a 'search', much as a solicitor might search for the title-deeds of a property that his client is purchasing. The citation of references validates many of the claims that he will make in his paper and embeds it in the pre-existing consensus. The orderliness of this process, the intellectual structure implicit in the library, the catalogue, the index, the encyclopedia, the treatise, give meaning to the research of the past and motive for research in the future. The mere accumulation of miscellaneous details is not enough to provide such order and meaning.

Yet the invention of a mechanism by which the results of detailed researches can be published piecemeal may have been a decisive step in the development of the 'Scientific Method'. Until about the middle of the seventeenth century, the communication of scientific information from one investigator to another depended upon private correspondence and upon the publication of occasional books and pamphlets. We really know very little about the network of private letters that passed from scholar to scholar in earlier epochs. The news of important discoveries might travel fast, but there must have been long delays between the claim to a result (e.g., a letter from Newton to Collins) and the definitive account in a public form. A private letter to a particular correspondent is an altogether different class of document from one that is to be published in a newspaper. The author may know his friend only too well; sometimes he need only hint at an argument to convey its meaning; at other times, he may deliberately hide his technique, for fear of plagiarism. The function of a letter may be more to indicate something that one ought to try for oneself—a way of putting lenses together to make a telescope, say, or a trick for solving a mathematical equation—than as a firm and reliable account of an established scientific result. Publication enforces a much more careful and explicit style, suitable for a wider audience, and amenable to direct critical analysis.

On the other hand, the separately printed and published pamphlet or book could easily get lost on its way from the bookseller to its potential reader. In the absence of a recognized centre for such communications, the task of keeping oneself informed about all matter of small scientific advances must have been exceedingly laborious and chancy. No doubt one would come quite soon to hear of and read the larger, more pretentious works—but these, as always, are often great rag-bags of rubbish. Without an efficient mechanism for the dissemination of briefer, more modest publications, research with limited objectives was gravely restricted.

The invention, therefore, of the scientific journal is of far greater importance than the other activities of the Royal Societies and National Academies which began to publish this new form of literature. In the first instance, these societies were created to hold meetings, at which scientific problems were discussed, and experiments performed. It was natural enough to provide members with written accounts of these meetings, to refresh the memory or perhaps to inform those who could not be present. Country members (Newton at Cambridge, shall we say) wished to have their turn to put forward their views. Foreign correspondents might write to the Secretary, with exciting news of fresh discoveries. The very name of the most famous of such journals—the *Philosophical Transactions* of the Royal Society—indicates that it began as little more than the minutes of meetings, printed for the personal use of the Fellows; but it soon developed into a regular periodical publication, containing 'communications' on a variety of scientific topics, much as we know it today. It is extraordinary to consider that the general form of a scientific paper has changed less, in nearly 300 years, than any other class of literature except the bedroom farce.

The advantage of the regular journal is that it offers speed and permanence of publication for the results of a great many investigations which would not, separately, be of much significance, but which interact with one another, stimulate further work and form the vast bulk of detailed observations on which major scientific advances are built.

At the same time, the very existence of a journal implies a degree of sociability amongst those who subscribe to it. The hallmark of a new discipline is the establishment of a specialized journal catering to the scholarly needs of its exponents. It constitutes an act of solidarity and sodality, and polarizes the subject around it. In the first instance the editor may have some difficulty in filling each number, and may encourage his colleagues to express themselves in print. The regularization of publication, the pressures of periodical journalism, may have had some effect on Science and Scholarship, generating a new feeling that time and speed were important.

To talk of speed as an attribute of such an old-fashioned eotechnic activity as the printing, on a hand press with hand-set type, of a monthly or quarterly volume, to be distributed by sailing ship, barge and mail coach, may seem presumptuous. Nevertheless, by the standards of eighteenth- and nineteenth-century Science, this was fast enough for the whole scientific community to become aware of new discoveries, to criticise them, react to them and act upon them. The whole of Europe was essentially within the same time radius as the world today. Modern scientific journals suffer the same production delays in the hands of editors, referees and printers as did their primitive ancestors. The interval between the dispatch of a manuscript and its appearance in print is rarely less than four months, and many reputable journals run a year or more behind, whilst air mail is too expensive to get the bulk of scientific literature from Europe to Australia, say, without further delays of several months.

It is conventional to bewail these delays, and to ask for a decisive increase in the speed of publication of good science. 'Why should we have to wait six months for a paper to appear in the old *Transactions*. Let's start a new journal that could promise publication upon receipt of the manuscript. Why, with modern technology, using photo-offset-multicolour-lithographic-stenography on computer links we could have the full text of every new paper relevant to your discipline, as key-punched from your personal interest profile, awaiting you on your desk every morning. The conventional scientific journal is outmoded, and should be superseded by a modern method of information storage, processing and retrieval.'

No doubt! Yet experience and observation do not confirm the notion that absolute speed of communication is of such enormous importance in Science. It is trite to remark that human thought processes do not speed up in synchronism with the machines to which they are linked. A barrage of idle words and trivial ideas can batter down the attention one should be giving to a few clear, simple and important concepts. There is a reaction time for the assimilation of a new idea, and a relaxation time for the decay of an old one. For most people these times are measured in months and years, not in days or weeks. The speed of conventional scientific publication is quite consonant with this scale of time.

Indeed, it is known from an actual investigation* that the publication delay is only a small fraction of the time between the gestation of a scientific idea and its general recognition and incorporation in the consensus. There are many stages of hypothesis, experiments, testing, confirmation, preliminary accounts at meetings and seminars, before the writing up of the definitive paper; and many more stages of criticism, reconfirmation, citation and reviewing before the results can be accepted and put into the textbooks. To concentrate only on the gap between the written and the printed text is to fail to understand the communication process as a whole.

Good scientists know this: why do they play along with the gee-whiz kids? There are two reasons.

In the first place, they worry about priority, and fear to be scooped. As we have remarked this is legitimate, for priority, or at least independent originality, of discovery is almost the only title that a scholar has to his own work. Nevertheless, this does not really require great speed of publication, provided that the convention is observed of printing the date of receipt of the manuscript at the head of each paper. It is distressing when two different journals have very different publication delays, so that an article received later in one may be published earlier than its rival in the other, but scrupulosity on the part of the writers of books and reviews can maintain equity of title, and there are natural processes of competition between journals tending to even up the delay times. In my experience, one is much more liable to suffer the neglect of one's work by having it published in an obscure journal, in the wrong country, than from sheer time delays. It is true that

* W. D. Garvey and B. C. Griffith (1964) *Science* **146**, 1655.

there are a few sharks nosing around in Science, ready to snap up unpublished ideas and make them their own; but speed of publication only facilitates such knavery by allowing *them* to jump ahead into print.

The other reason is that beneath the surface layer of formal publication in science there exist many networks of informal communication. The old courtesies of private correspondence, the new vulgarities of conferences and meetings, interchange of manuscripts and data, sabbatical leaves, consulting visits, seminars, conversation round the coffee table—these knit together the scholarly world in a way that is scarcely evident to the outsider. These are the links that bind the members of the 'Invisible Colleges', who are conscious of working in the same field, as colleagues and rivals, throughout the world. Modern technology—the telephone, air mail, jet travel—has greatly speeded up these channels, which are now much faster than the official publication procedures. Having just attended a conference at which is announced the Non-Conservation of Parity, or the breaking of the Genetic Code, one is irritated at the idea of waiting six months for the published paper, with all the details. It seems absurd not to distribute, say, a duplicated verbatim transcript of the great man's lecture, or a quick, if crude, 'preprint' of the paper he proposes to publish.

There are, indeed, occasions when scientific discoveries of great importance need rapid publication, and mechanisms have always existed for this purpose. Thus, in the late nineteenth century a 'letter' to *Nature* was the accepted medium for such communications, and would be printed and published with due urgency. But this urgency does not apply to the vast bulk of scientific literature. Most scientific papers are in fact, read by very few people. It is only in special moments of 'revolution' and 'crisis' that large numbers of scientists are working on the same problem, and need to be rapidly informed of any new development that might be of decisive influence in their own work. For most purposes, the unofficial channels are quite adequate as a grapevine of hints and ideas, observations and opinions.

It is often argued, with some justification, that this is all right for established scientists and their students and colleagues, who belong as of right to the appropriate Invisible College, but that it is difficult for outsiders and newcomers to break into such circles and find out what is going on. This is true, up to a point, although the proliferation of conferences at which personal contacts may be established makes it less compelling. The pace of Science in fashionable subjects does sometimes make it difficult to determine, from the published literature, what is in fact going on, and one often observes that work in countries such as India, whose scientists have little opportunity to travel, is out of touch and out of date. Nevertheless, it is important to distinguish between lack of contact and lack of comprehension; the time lag between the acceptance of a new idea in the best and worst laboratories in a given country may be just as long. Merely being at the receiving end of a rapid transit system for current publications is not enough to make one master of it all. There is no substitute for a long apprenticeship in a discipline, and years of hard work trying to understand what it is really about.

The real issue is that the distinction between formal and informal scientific communications should not be blurred. The official scientific paper in a reputable journal is not an advertisement, or a news item; it is a contribution to the consensus of public knowledge. As we have already seen (chapter 3), it is written in a special impersonal form, in somewhat abstract language, within a strong convention of form and style. It must claim no more than it can substantiate, it must not criticize other work needlessly, it must give all due deference to previous work on which it depends, and so on. A major achievement of our civilization is the creation of this form of communication, however clumsy and barbaric it may seem to those whose concerns are with poetry and feeling. The individual primary paper is not the final form of the consensus but it is the brick from which the whole edifice is to be built.

Word-of-mouth communication can never conform to these conventions. The choice of language is inevitably haphazard and elusive, and is always accompanied by gestures to convey the meaning. To publish verbatim the 'discussion' on a paper at a conference is to give unwarranted permanence and solidity to the contingent and transitory. Sometimes the notes of a lecture can be of value as a port of entry to some difficult intellectual terrain; this value may be destroyed by the attempt to make the language official, suppressing the 'hand-waving' that is so important in carrying intuitive meanings.

A letter, whether private or 'to the Editor', is also no substitute for the full formal account of work done. Brevity implies compression of the argument, suppression of important steps and the omission of *caveats* modifying and moderating the major point at issue. Sometimes a letter is just a very short paper, to which there is no objection; all too often it is a claim to discovery substantiated only by the scientific standing of the author. The whole scientific method is travestied when the author announces 'I have shown that...', instead of saying 'If A, then B, hence C: do you not agree?' My own experience with 'letters' journals (which are now quite common) is that they contain a great deal of rubbish—either trivia, or false claims—with very few articles that merit the privilege of rapid publication and careful attention.

Another mechanism for the dissemination of scientific knowledge in advance of formal publication is the distribution of 'preprints'. For the reader who is unfamiliar with this practice let me explain that a preprint is a clumsy, bulky, semi-legible document, being a duplicated version of a paper submitted for publication but not yet accepted and printed. It is a mechanised version of the decent and proper custom of writing to one's friends, colleagues and rivals about one's current work, with perhaps a carbon copy of the manuscript or a galley proof as a special favour. Perhaps this custom is harmless enough, although it always seems grossly extravagant in duplicating costs and postage. Preprints are unpleasant to read (especially if the mathematics is clumsily typed or written on the stencil) and most of them that one gets are quite irrelevant to one's interests, being distributed according to some vague list in which one's name appears because one published a note on the subject ten years before, or because one is held to be an authority on the field in general. It is seldom that one in fact wants to read it at all, and one can well wait until the paper gets published. Only occasionally, in a closely knit and active field, do preprints keep the Fellows of the Invisible College in closer touch than can be achieved by personal letters and contacts.

The danger is that attempts will be made (indeed are being made) to formalize and centralize the distribution of preprints to make it 'efficient'. For example, the U.S. National Institute of Health recently experimented with machinery for the duplication of good typescripts, for mailing to selected lists of subscribers and correspondents in the various fields of biology for which it is responsible.

At this point, we step over the line between informal and formal dissemination of information, between the private communication and the published paper. The objection is not that those who receive such preprints are especially favoured; it is obviously necessary to allow anyone to apply to be put on the mailing list if he has minimal scientific qualifications. But then the document has become 'public'; from the point of view of the reader it is no different from a paper that he might obtain by borrowing or purchasing a copy of a regular journal. He will feel at liberty to refer to it, quote from it and rely upon its authenticity, just as if it were a part of the official literature.

Why object so vehemently to such a harmless, if extravagant, custom? Surely one ought to allow freedom of publication in Science, which is so committed to the principles and norms of freedom of speech and comment.

The fact is that the publication of scientific papers is by no means unconstrained. An article in a reputable journal does not merely represent the opinions of its author; it bears the *imprimatur* of scientific authenticity, as given to it by the editor and the referees whom he may have consulted.

The referee is the lynchpin about which the whole business of Science is pivoted. His job is simply to report, as an expert, on the value of a paper submitted to a journal. He must say whether the results claimed are of scientific interest, whether they are authenticated and made credible by sound experimental methods and good logic, whether the paper is well expressed, not too cryptic nor too verbose, with adequate references, etc. He reports to the editor, and although this report may be sent to the author for action or rebuttal, he is protected by anonymity.

To referee a paper is thus a serious responsibility (although it does not expose one to the importuning and recrimination that fall upon an editor). It is not at all easy to read a paper on a topic with which one is not totally familiar (for there cannot be such an authority on every *new* point of science), and to decide whether it is of the required standard. Sometimes one may pick out an obvious technical flaw; but the grounds for rejection are usually much less definite. One may decide, for example, that although the experimental results are valid their comparison with theory uses outmoded concepts which no longer make them interesting. Or one may have to report that the calculations, although correct, are based upon assumptions that are not very close to reality, and hence are irrelevant. Or it may be that the whole paper is so clumsily written that one can scarcely puzzle out its meaning, which then turns out to be extremely interesting; should it be sent back for the long process of revision, to an author who is obviously ill-equipped for such a task? If it is incomprehensible is one justified in rejecting it out of hand, at the risk of it being discovered later to be a work of genius? How far should one go in sustaining a controversy with an author, who refuses to modify his claims on some key point of his argument? It is a thankless task, whose pleasures are rare and brief ('This is an excellent paper, which I much enjoyed reading; it should be published at once'), or reprehensibly sadistic ('The author might have considered the following objections, which reduce his whole argument to mere speculation (1)...').

Are not the powers of the referee too great? Is he not that monster, a censor? First, remember that there are safeguards against abuse. It is usual to consult two or more referees, and a paper is only rejected if they both, independently and decisively, recommend against it. If there is conflict of opinion, then further referees may be consulted, and the whole matter will be given careful consideration by the editorial board, with always a prejudice in favour of publication.

Again, the business of refereeing is spread over the whole body of active scientists in the particular field covered by the journal. It is true that the more senior persons may be consulted more often, but they have no exclusive privilege in the matter, and are always happier to let the work be done by their junior colleagues. They are not, therefore, obstinate tyrants, to be placated; although each referee may have his own personal standards as an author, he usually accepts the general criteria of the journal to which he is reporting. A consensus as to what constitutes 'publishable' work is established, and referees who are persistently too harsh are dropped from the list.

In any case rejection by one journal does not prevent publication by another. There is no general conspiracy in the scientific world to suppress the works of particular authors. Although there may be some rationalization of journals within a particular country in a particular discipline, there is still plenty of competition between one country and another, and between commercial publishers and learned societies. A referee is always conscious of the danger of rejecting a brilliant work through failure to understand it. There are several such episodes in the history of science, such as Waterston's paper on the Kinetic Theory of Gases, which was refused publication by the Royal Society in 1845, although it anticipated Joule's work by 20 years. Perhaps Waterston's referee was not altogether to be blamed, for the paper is cryptic and obscure; but history is cruel in retrospect to such a blunder. We may comfort ourselves that there are many alternative outlets nowadays for such work, which would probably not now be condemned to the obscurity of the library at Burlington House, where it was discovered half a century later by Lord Rayleigh.

Or are we being too complacent? Is there a hidden treasure of rejected works of genius which would have revolutionized our view of Nature had they been published? I doubt it. The sort of work that gets privately printed and distributed to professional scientists in a desperate attempt to catch their attention is almost always quite worthless. In Physics it is usually a general theory of Space, Time, Matter and Radiation—an attempt to outdo Quantum Theory and Relativity, Cosmology and the Theory of Elementary Particles in one splendid stroke. The very form of such literature exemplifies my general thesis. Superficially, the author has a logical argument, a rationale, an interpretation of everything. To the layman it may seem no less compelling than Einstein's paper on Special Relativity, or Bohr's work on the Quantum Theory. But by his contempt for the current consensus, by his condemnation of all accepted theories and his insistence that he alone is favoured with the true light, the author puts himself outside of the scientfic community, and beyond serious scholarly consideration. Yet, God knows, one of these cranks might occasionally hit upon a Truth; that is for one or two of us perhaps to discover accidentally for ourselves by listening or reading occasionally; unless he is willing to conform to the norms of Science, he cannot expect an immediate entry into scientific circles. In this aspect, professional science is, indeed, largely a closed, self-validating system, and perhaps subject to dangers that we do not recognize.

Perhaps the most dangerous post for a referee or editor is at the boundary of a well-established discipline where a new subject is striving to be born—such as, for example, orthodox Medicine at the moment when Freud began his clinical studies in Psychiatry. In its initial phase the new subject lacks a consensus to which new ideas may be referred. From the point of view of the older discipline, the attempts to create such a consensus seem wildy speculative, controversial and contradictory—in short 'unscientific'. To admit this sort of work into the literature seems to imply a general lowering of critical standards, laying one's cherished field open to cranks and lunatics. It takes a stronger imagination than is usually found amongst scholars to recognize the validity and importance of such writings, and peculiar intelligence and foresight is required to distinguish the reports of serious exploration from mere travellers' tales. The wisdom of hindsight is not kind to the scandalous cases where orthodoxy and 'subject snobbery' have hindered the growth of knowledge; nevertheless, we must be sympathetic to the dilemma of the honest guardian of scholarly standards in these perplexing circumstances. Cosmology, for example, is a speculative subject, where the referees have learnt not to be too tough!

In 'normal' science, however, the referee must not be too lax and kind. Once a paper has been published, it will probably not receive much further explicit criticism. The courtesies of high Science forbid unnecessary controversy. If the work is wrong, it is more likely to be ignored than deliberately gainsaid. It is thought to be unkind and professionally unprofitable to point out another man's error in public; better simply to hint that his claims are not quite substantiated and then to give one's own more positive and correct version. The referee is not to be expected to prevent every error; but he can ensure that the argument of a paper is worth attending to. He does not, like a moral censor, suppress improper sentiments; but he does insist that new ideas that might shake the current consensus be expressed as accurately, clearly and plausibly as the author can make them. He is, so to speak, a traffic policeman on point duty, keeping the traffic flowing smoothly by imposing an orderly succession and conformity to the general rules. To do without him is to invite chaos. There would be no check on the style of individual papers, and no means of distinguishing a few genuine results from a mass of rubbish.

The experienced professional scientist seldom comes into conflict with the referees of his papers, not because he belongs to an inner conspiracy of mutual admiration but because, as I have indicated, he has internalized the standards that the referee is trying to enforce, and has already anticipated most reasonable grounds for criticism; in other words, he has already learnt to drive with due care and attention. That is why a few journals that do not referee their contributions are not actually completely disreputable; by accepting work only from recognized professional scientists they can rely upon the personal standards of their authors. But there is no guarantee that the whole scientific community could maintain this sort of intellectual integrity if it were not for the referee system. To point out that, say, only twenty per cent of manuscripts are rejected by the referees does not prove that they are unnecessary; it shows that most authors have taken sufficient care with their work to satisfy the standards which the referees are designed to enforce. The 'scientific attitude' is not an inborn virtue; it is a reflection of the intellectual norms imposed by the community to which the scientist would belong. An attack on the function of the referee, whether by the pseudo-publication of 'preprints' or by the actual publication of unrefereed 'letters' journals and such like rubbish, is a blow at the roots of Science itself.

To the reader unfamiliar with this system, the picture I have sketched in this chapter must seem either a travesty of their ideal of Science as intellectual liberty, or else a confirmation of their suspicion of its inhuman and mechanical dogmatism. Let me try to remedy these doubts and fears.

Remember that the 'official' scientific literature is not the only path along which intellectual ideas may pass. The whole ideology of Science, the principle of a *freely accepted* consensus, implies a society in which there is general freedom of speech and comment.* As I have indicated, the *informal* system of scientific communication is quite as important as the formal system, although having a different function. When editors and scientific societies have blocked the publication of unorthodox ideas in their journals, they have seldom been guilty of sins worse than those of the clique or coterie. The crimes of Lysenko were only possible in a country where *all* channels of communication could be denied to his critics.

On the other hand, liberty of the individual is only guaranteed by orderly institutions. The delicate tension between imagination and criticism that drives Science intellectually must have its counterparts in the social sphere. The forces of professional competition and advancement, of corporate and national rivalry, of economic development, of the sheer growth in the size of Science itself, are all antipathetic to the autonomy of the scientific community as such. As I shall emphasize in the next chapter, the professional

* This is particularly well stressed in Polanyi's writings.

associations and societies are not very significant as such, and do not constitute a countervailing power. It would be a far more serious threat to intellectual liberalism if, following some ominous trends in the Ph.D. system, they were allowed to close their ranks by assuming power to licence or register 'practising' scientists—i.e. those permitted to publish—like doctors or lawyers.

The only strong and well-supported institutions of the scientific community are its professional journals. Concentrating solely upon the plausibility and relevance of the arguments advanced in a paper, without deference to the identity of the author or his corporate backing, the referee system provides the only protection of scientific standards against the demands for quantity of publication without regard to quality. It is proper that the scientific world should be polarized socially about its intellectual pursuits, and should be concerned mainly with the validity of *ideas* rather than the accrediting of persons. The integrity of its scholarly publications must be its citadel.

There are, as I have suggested, some signs that this basic principle is not well understood. For example, a system has grown up in recent years by which a journal levies a charge of so many dollars per page from the institution employing the author of a published paper. In practice, this is no great hardship, for it is paid out of the government or corporate funds supporting the research that is being reported, and it allows the journal to be bought much more cheaply by its readers. I am informed that the decision of the referees on the paper is quite independent of whether this 'page charge' is honoured, and that the integrity and standards of the journal are preserved. Nevertheless, in principle, for anyone who is brought up to believe in the real power of monetary transactions, this looks perilously close to advertising. It seems to shift the responsibility for the validity of what appears in the journal away from the editor and referees, as representatives of the readers, to the corporate employers of the authors. I do not point to this phenomenon as a symptom of present corruption, but to indicate the absence of a clear grasp of the nature of Science itself in those high circles where this system was devised.

One thing that referees or editors do not succeed in improving is the literary style of most scientific papers. This is a topic of perennial complaint in scholarly circles. As I have already remarked, the impersonal phraseology is an attempt to make the work seem already part of the consensus. I think this is why vivid phrases and literary elegances are frowned upon; they smell of bogus rhetoric, or an appeal to the emotions rather than to reason. Public Knowledge can make its way in the world in sober puritan garb; it needs no peacock feathers to cut a dash in.

Unfortunately, the ideal of good plain prose is seldom achieved. In the first place, scientific arguments are often exceedingly subtle and difficult in themselves. The convention that it be expressed in linear prose, rather than symbolically or diagrammatically,* can turn a relatively simply piece of logic into something resembling a legal contract. The drafting and re-drafting of such material calls for more training, experience and natural talent than is usually given to the ordinary scientific research worker.

The attempt to write each paper as if it were already accepted also causes the author to imitate the style of other accredited papers. He uses the standard technical words and phrases of the subject, not out of any desire to baffle the layman, or to associate himself prestigiously with his would-be colleagues, but because these seem the natural language of the intellectual territory of his argument. He knows that he is writing for experts, who have themselves already expressed their views in this technical language, and who therefore may be assumed to understand it. Of course it often happens that words are reified, and the arbitrary classifications and distinctions implied by a jargon become frozen into over-mighty systems of thought—but these are the risks of all serious discourse. The esotericism of scientific writing springs more from laziness, and the modesty of such imitative attitudes, than from arrogance or the desire to dazzle.

Ordinary papers are sometimes called the 'primary' literature of Science. They are the permanent records of research done, and

* Meredith's book is an eloquent plea for a new diagrammatic language for the communication of science.

are stored in libraries as bound volumes of journals. They constitute the official archives of the scientific community. The job of a research worker is to produce further 'publications' to add to these archives. Some scientists regard this as their sole duty.

But an archive is useless without a catalogue or index. The bulk of scientific matter now being published is enormous—something like a million papers a year, in tens of thousands of journals, in the conventional pure sciences. Storage space for these vast numbers of documents can be constructed; but how is one to have intellectual access to them? If the knowledge that they contain is to be made the basis of use, or of further knowledge, how are we to discover when we need it: obviously, we cannot read all these papers ourselves, and remember what was in each one.

There are, of course, certain ways in which papers in various fields are fractionated and concentrated for our specific attention. Scientific journals become very specialized; we do not read a journal entitled *All of Physics* but subscribe to *The Physics and Chemistry of Solids* or *The Journal of Scientific Instruments* or *The Journal of Fluid Mechanics*, each of which may contain a high proportion of papers relevant to our interest. There are good journals and bad ones, so that we only keep in touch with those likely to contain good papers by reputable authors. There is a sort of law of specialization that ensures that the number of journals containing significant papers on one particular subject does not increase; one simply narrows one's field of vision to keep the influx constant.

Nevertheless, one cannot rely upon one's memory or personal card index outside of such a narrow field. There exist organizations whose function is to provide indexes to the whole of the official literature of the major sciences. The conventional method is for the author of a paper to provide a short 'Abstract' or summary of his work, which is printed with the paper for the benefit of the reader. These abstracts are then collected and reprinted in special journals, where they may be referred to by means of various kinds of index, by subject, by author, etc. Thus, if one wishes to find out whether the electrical conductivity of copper has been measured recently, one looks in the latest volume of *Physics Abstracts* under the heading 'Copper: electrical transport properties of', where there is a reference to abstract no. 32123, which turns out to be an account of a theoretical calculation of the thermoelectric power of copper in a high magnetic field, by X. Y. Z. Smith and A. B. C. Jones, and not what one was after at all. So one then begins a more systematic search in earlier volumes, under various possible index entries, until one finds an abstract that sounds relevant. One must then chase around the library for, say, volume XIV of the *Science Reports of Tohoku University* (which may not, of course, be available), where the original article is to found. If one is lucky, the result of this search is mildly useful.

'Searching the literature' is thus a slow, tiresome and uncertain activity; it may yield little fruit, and one never has the assurance that all the relevant papers have in fact been found. It is not surprising that this apparently mechanical procedure has attracted the attention of the computer engineers, who have proposed to automate it for us. Their ideal system of 'information retrieval' is a typewriter in every scholar's office, where he may tap out a request for all that is known about the electrical conductivity of copper, and receive on a television screen facsimilies of the appropriate papers. Rudimentary systems of this sort have in fact been constructed, and can be made to work more or less as intended.

But I do not propose here to discuss the problems involved in making and using such devices. The point is that the published literature is being treated as an archive of what is 'known' scientifically. But what we retrieve from this archive is 'information', or 'data'—it is not 'knowledge'. The myriads of primary scientfic papers, with all their facts and opinions, calculations and hypotheses, are the material out of which the consensus is to be constructed, but they do not explicitly constitute this consensus. To concentrate solely upon the 'retrieval' of the primary literature of science is to miss a most important point of principle and practice. A scientific paper is not to be validated on its own terms. Even after it has been refereed, the 'information' that it proffers is no better than the competence of the author and the consistency of his arguments and experiments. An inhuman style does not guarantee superhuman truth.

Following the publication of a paper, there is a slow, unconscious and very erratic process of evaluation. Of course, an important paper announcing a major discovery in an active field is quickly criticized, and either written into the canon or ruthlessly rejected. People will try to repeat the experiments or go through the algebra to get the same results (sometimes they fail: then what? Is somebody lying? Is Nature capricious? More experiments and even more confusions!). But even this may take several years; one is often wise to be sceptical of new results that seem completely at variance with accepted opinions, until they have been confirmed independently; and one is *always* wise not to announce such results until all possible avenues of doubt have been closed. To 'scoop' a bogus 'discovery' is the road to scientific Siberia.

One sometimes finds a succession of papers on the same topic, each referring back to and correcting the errors of its predecessors—for example, pointing out systematic errors in an experimental measurement, and proposing to avoid them by a new method. One tends to treat the latest of such a chain as the most reliable, although, in fact, this is not always warranted. A new method may after all introduce a new systematic error, which the author has not bargained for. To suppose that the most recent work, with all the benefits of earlier papers to refer to and new techniques to be used, is necessarily the best is to forget that sometimes a young man of today is rather more stupid than one or two of his predecessors.

Most scientific papers in the first instance are taken more or less at face value; one does not rely strongly on them, but one does not go to the trouble of confirming them in detail. One supposes that the author knew his job: if the results are not surprising, that merely confirms one's trust in the uniformity of Nature. By a process of passive acquiescence, the paper gradually becomes part of the consensus.

What I wish to emphasize is that this sort of unsystematic

weighing up of published scientific work is sceptical and cautious, and very far from the positivist moods of 'verification', and 'falsification'. Experience teaches one not to believe the evidence of another person's eyes; it is best to wait for the testimony of several witnesses before giving credence to new ideas. The political radicalism of some excellent scientists reflects their energetic imaginations; the deep conservatism of some equally good ones is rooted in this scepticism and caution. They do not so much wish to defend an orthodoxy as to ensure that additions and modifications to the consensus shall be of the same indisputable veracity as its firmest foundations. There is little comfort in forsaking an island of rationality—however threatened by possible disaster—for one that is not obviously better supported.

A scientific paper moves out of this limbo of tacit acceptance when it is referred to in a 'review article'. The function of this type of 'secondary' publication is to give an explicit account of the current consensus in some particular field. The author is expected to read all the papers on the subject, give a brief account of their findings, and relate them one to another, noting agreement and contradictions. From such a review, one can learn the general credibility status of each paper, and the weight of its contribution to the advancement of the subject. It is mandatory to cite every relevant paper, but the courtesies forbid that errors should be bluntly confounded. Indeed, the language of review articles has its own conventions. 'Wrobinson (1948) has shown that...but this is not confirmed by the work of Qwhite (1951), who obtains ...'; 'As Hblack (1957) pointed out, the incoherent scattering may possibly have contributed an extra term which would explain the discrepance between theory and experiment observed by Kgreen (1954)'; 'Gknight (1960) has made a careful analysis of the 57th virial coefficient, which turns out to be the order of 10^{-16}; on the other hand Bpotts (1959) has shown that the numerical factor in the first term of the series may be in error by as much as 10 per cent'—all of which, interpreted into the vernacular, just mean 'Damn fool!', 'Idiot!', 'Nincompoop!'. The writing of a review article so as to do justice to the truth whilst preserving the friendship of one's colleagues requires a good deal of tact and art, and most of us baulk at the stiffer jumps. It is best simply to ignore folly completely than to attempt to expose it; and are we not all vulnerable in this respect?

The tentativeness of the assessment of recent work in most review articles indicates that the current scientific consensus is always fluid. One may say, indeed, that the only consensus is that some new theory has not yet become part of established knowledge! But there comes a time in the history of a discovery or hypothesis when it is finally accepted or rejected by the scientific community. This may take a long time, as with the theory of Continental Drift, or it may occur almost overnight, as in the theory of the Non-conservation of Parity in Weak Interactions.

In my view the gravest weakness in the organization of modern science is the lack of systematic exposition of the consensus at the stage between the review article and the undergraduate textbook. There is a peculiar gap between the *catalogue raisonné* of the typical *Recent Advances in*... article, and the dogmatic authority of the lecture course. Experimental practitioners of any field of research will tell you what ideas are well understood, and accepted by everyone, and what is still speculative and uncertain, but there is a reluctance to set this down on paper. In principle, this is the function of the treatise or monograph; but all too often such volumes are mere compendia of review articles, which do not give judicious general account of the subject as it exists.

The importance of such a general account is that it turns *information* into *knowledge*. The aim of Science is understanding, not the accumulation of data and formulae. We need to bite into whole concepts, not swallow them piecemeal. To fail to construct the building, to leave all the bricks lying round in untidy piles, is the *trahison des clercs* of today. The separate pieces of information in the separate primary papers need to be joined to one another, fused with one another, welded into a coherent intellectual machine, which may be used as a whole, whether for material benefit or for further scientific exploration. The work of synthesis is quite as important as the analytical process of discovery and experiment and must not be left to unconscious forces, or to the accidental requirements of a graduate programme.

As I see it, the duty of every active scientist requires him to stand back from his current research from time to time, to take a look at his subject as a whole. He ought to ask himself 'What major question are we asking of Nature? How far have we succeeded in answering it? What are the assumptions upon which our research has proceeded? Are all the phenomena subservient to an intelligible theory, or are there major mysteries and contradictions? How does what we know in this field connect with our knowledge in other fields? Have we concentrated too much on the simple cases, and failed to attack more complicated systems which may be of practical importance? Is our solution of an idealized model problem relevant to the real solution? Was the labour of elaborate solutions of idealized models worth more than the fun of puzzle-solving? In short: what do we really *know* about our subject?'

It is interesting to note that books attempting to answer such questions come to dominate their fields for a whole generation. They are rare, so that their authority stands almost without question: they are accepted and cited unthinkingly for years on end. Yet the true consensus is continually growing and changing. In quite ordinary fields of 'normal' science, an interval of five years may produce quite new insights, quite new modes of comprehension and logical connection, which do not 'revolutionize' the subject, but which are as important as major new discoveries in making the pattern of things clear. Indeed, it would be wrong to suppose that there could be, even at this stage, a single authentic point of view. The active research worker needs the benefit of several different attitudes to, and aspects of, the subject of his devotions. Perhaps the most beautiful discovery in the whole of quantum theory was the reconciliation of Heisenberg's Matrix Mechanics with Schrödinger's Wave Mechanics—the proof by Dirac that these were equivalent mathematical systems. It is only by making apparently contradictory systems of ideas explicit that one may show the connection between them.

The oldest principle of the Philosophy of Science is Occam's Razor—the hypothesis of maximum simplicity. If knowledge is to be useful, it must be simple; it must subserve the greatest variety of phenomena to the most economical set of laws. It must surely be our object, in collaborating in search of consensible knowledge, to transform this into the best possible, most easily comprehensible pattern of ideas. The case against vast capital investment in systems of 'information retrieval' is that it implies that the best deployment of the human resources of Science is to put every man with rifle and bayonet in the front line, leaving staff work and strategic command to mindless automata. To neglect the writing of monographs and treatises, to treat this as somehow not 'research work' like watching meters and scribbling algebra—something to be done in the evenings, in one's 'own time', as a supplement to one's income—is to betray the scientific tradition. To find something out, look first in a book, not the abstracts!

It is commonplace to remark that science has become remote from the layman, that no one condescends to explain it in simple language, that it is an esoteric cult. I do not believe that the need is so much for the 'popularization' of Science for the general public as for the popularization of Science for scientists themselves. The information system of Science works quite well for the accumulation of the details, but is failing in the equally essential task of assembling these details into a comprehensible, analytically ordered and coherent system of ideas—a consensus—public knowledge.

In this chapter I have discussed some of the issues that have prompted many scientists to assert that the whole communication system of Science is in a crisis. Bernal* expressed this opinion before the Second World War; the difficulties to which he pointed have been aggravated by the continued growth of Science, and by the invention of wholly new techniques for the duplication, transmission, storage and retrieval of documents. I am not sure that I agree that the supposed crisis is so serious as is made out—for

* J. D. Bernal, *The Social Function of Science* (London: Routledge 1939).

example, much of the demand for 'instant' publication and information retrieval seems to stem from a neurotic obsession with speed in highly competitive and fashionable fields than on a calm assessment of the actual needs of regular research.

But this is not the place for a three-decker sally, broadsides booming and marines at the ready, into this controversial topic. What I have tried to do is to show that a view of Science comprehending its personal, intellectual and social aspects helps one to understand the issues more clearly, and suggests some principles on which existing and proposed practices may be judged. If I have sometimes engaged in polemical skirmishes, this is again to indicate that the consensus principle is no abstract formula, but provides a philosophical basis for action.

7

INSTITUTIONS AND AUTHORITIES

'The question is', said Humpty Dumpty, 'which is to be master—that's all'. LEWIS CARROLL

Most modern scientists are the employees of formal bureaucratic organizations—universities, civil services, industrial corporations. They are paid to do their research at rates decided by quite conventional mechanisms, involving, in many cases, collective bargaining, Treasury sanction, the forces of the market, scarcity value, etc. They are often accorded status and official power over their colleagues, as chairman of department, research director, professor, reader, and so on. From the outside, they seem to be far more bureaucratized, far more like officers in so many little armies, than many other professional workers, such as doctors, lawyers or architects, who often still perform their duties as individuals in private practice.

Yet this impression is deceptive. Once a scientist has passed beyond his doctoral apprenticeship, he is given great freedom in the choice and manner of his research. His hours of work are seldom prescribed; he usually claims immunity from 'clocking in'; and he may spend a great deal of his day nominally at work yet apparently just gossiping, drinking coffee or staring out of the window. Within the boundaries permitted by his employment, when he has completed his modest stint of teaching, or as an 'administrator', he feels himself to be supremely his own master, defiantly 'self-employed'.

This is, of course, as it should be. Nobody who has the least understanding of the demands of the intellectual life insists that it be ordered to a formal pattern. These demands are generated internally, and do not, in the first instance, meet with the lives of other people. A shop assistant must be at his counter, ready for the customer whom he is to serve. The doctor must attend to his clinics and be at the beck and call of patients in their homes. The research scientist wants no clients, nor interruptions to his own private dealings with Nature. If he is employed as a teacher, or as a technologist, he seeks to arrange his day or his week so as to keep as much of it free as possible for what he (revealingly) calls 'his own work'.

The whole question of the allegiance of the scientist to the organization that employs him has been discussed at length in the literature on the management of industrial research.* My impression is that the sort of scientist with which we are mainly concerned in this book—that is, more a 'pure' scientist than a technologist—often feels no more than cupboard love for the organization for which he ostensibly works. He regards it as a convenient habitation, a source of income, a landscape within which to build his own personal, private life. No doubt such organizations are necessary, and have to be governed, but he would rather other people took on these responsibilites. Of course he wants lots of money for his apparatus and may learn to become very cunning and selfish in special pleading for it, but the major purposes for which the great corporations exist—education, defence, profitable production, national prestige—may be of little moment to him. If a radar research laboratory, devoted to the development of military technology, happens to be the best place he can find for his study of compound semi-conductors then he will be quite happy to have a niche there, feeling virtuous in the thought that 'wicked' defence spending is being used to support such 'good', ploughshare-worthy activities as his own researches.

Of course, this mood, or ideal of 'dedication to Science' is typical of the man fresh from graduate school, and does not always last. Many discover the fascination of technological development, the joys of management and corporate intrigue, the satisfactions of teaching, the responsibilities of a human task to be done. Many scientists do, in fact, become administrative civil servants, executives of businesses, Presidents of Universities. But they do so

* E.g. W. Kornhauser, *Scientists in Industry* (Berkeley: University of California Press, 1962).

with mixed feelings, knowing that it will be said of them sadly that 'He's only an administrator nowadays', or 'He's quite given up research'. To the ordinary middle-class man of moderate ambition, 'getting on the Board' seems both a reward and an opportunity; to the research scientist it may seem almost like a betrayal.

This indifference to institutional goals, this unwillingness to take administrative responsibility, has often been commented on by students of the sociology of organizations. It is the key problem in making scientific research workers amenable to the demands of productive industry. Even in universities, the conflict between research and teaching, the reluctance of scientists to pull their weight on faculty committees, the attraction of a pure research Chair, without other burdens and responsibilities, are symptoms of the same disease.

It is not as if their professional associations were more significant to them. There are, of course, numerous societies, covering the major scientific disciplines, Institutes of Physics and Chemistry, of Biochemistry and Zoology, of Astronomy and Anthropology, and all the 'ologies of the modern scholarly world. They organize conferences, publish journals, pass resolutions, award prizes, discuss questions of education and professional ethics in a thoroughly praiseworthy manner. But these associations are usually run by very small groups, not quite cliques, but seldom covering the vast mass of their members, who, like ordinary trade unionists, are perfectly happy to have these useful activities governed more or less efficiently by those few persons in the profession who really care about such things. The average scientist regards the meeting organized by his Association, its journals and other publications, as useful adjuncts to his work, which might have to be invented if they were not already in existence, but no more positively and passionately than the buildings of his laboratory or the equipment in its stores. He will agree to serve as an expert on a committee, or to act as a referee, or perhaps, under duress, to edit a journal, but with no deep commitment to the association as such.

Why is it that active research scientists are so often lacking in zeal on behalf of their official institutions, even when these exist for the furtherance of pure learning and other admirable ends? Why are they so contemptuous of the honourable, necessary and often exciting work of management and administrative authority?

For a few scholars the pursuit of knowledge is itself so all-absorbing and entrancing that they become completely drugged by it, and sacrifice their families and their colleagues to it. For a few, the life of the hermit saint is fully satisfying; they go through life as in a dream, ministered to by a sensible and devoted wife, treated gently, as a valuable mascot by their academic confrères. Art is indeed long, and life is short and they follow their star.

But the typical professional of whom I am generally writing is not usually more 'dedicated' to his job than his lawyer, engineer or business contemporary. He believes in working hard and getting on, and he is not usually so bohemian as his humanist colleague. In their home lives, scientists are good husbands and fathers, frugal of their resources, ready to turn a hand to lawnmowing, barbecuing, house painting, car washing, the general needs of the community, and so on, like any respectable middle-class sales manager; yet when they get to the laboratory in the morning, they claim the right to shut their office doors against such ordinary necessary tasks as preparing next term's lecture list or interviewing a new batch of students. Even when tenure has been achieved, and there is no promotion to strive for, doing research, 'doing one's own work', is still the first priority.

* * *

The fact is that the scientist gives his allegiance to the scientific community, in particular to the 'Invisible College' of his specialist field of study. His true loyalty is to the informal institutions that underly and maintain the search for consensible knowledge. The true sociology of Science is not concerned with the relationships between Science and Politics, or scientists and politicians, between Science and Industry, or scientists and industrial corporations, but with the social interactions between a scientist and his colleagues—those other scientists studying the same problems, whether in Europe, America or Timbuctoo.

The point is now well taken, and studies exist of some features of these tenuous tribes.* But the whole subject is of extraordinary difficulty and delicacy. In the first place, the only thing that the members of such a group have in common is their interest in a non-material, intellectual matter. They are bound to one another neither geographically nor fiscally. There are no written rules of association, nor firm procedures for the enforcement of discipline. There is no certificate of membership, nor are there clear hierarchies and grades of status within the group. I do not believe that even the best attested principles of anthropology and sociology must necessarily apply to such institutions, which are at the same time voluntary and highly self-aware, yet so vague and unsubstantial. It is certainly essential to have a very good understanding of the nature of Science itself—the subject matter of the present essay—if one is to make any progress in this subtle subject.

How, for example, does one become accepted as a charter member of an Invisible College? It is not sufficient merely to be doing research in that field, or to publish a paper on it, unless this is of such extraordinary merit as to catch the attention of the whole community. The usual entry is achieved by patronage. The students of a particular professor are recommended to his colleagues not merely for jobs, but as potentially able contributors to the field. To have taken one's doctorate in some famous school of research provides one with a ticket of admission, as a visitor or temporary research worker, to another distinguished group, where one's name will then become known. A joint paper with a leading scholar may be mainly the work of the student—but the name upon it may provide the necessary *cachet* for his further advancement.

It is very important to meet people, and be met by them. One of the main functions of colloquia and seminars, where distinguished speakers are invited from other universities to talk about their work, is to clothe abstract personal names with flesh and blood, so that the young research worker may become familiar with the senior members of the College to which he will belong, and may even have a chance to be introduced, and talk informally

* Storer, *The Social System of Science*; Hagstrom, *The Scientific Community*.

about his own research work, so that he will be remembered and recognized in future.

For those who lack a powerful patron, an obvious move is to give a 'contributed' paper at a conference. This is a rather unsatisfactory business, which most scientists complain about, but most will not forgo. The usual programme at a meeting of this sort is to have a few long 'invited' papers, by acknowledged experts, to survey the field or to announce their latest discoveries, together with numerous short contributions offered by any member of the general audience. It is, of course, almost impossible to say anything comprehensible or significant in about ten minutes, but there are always more papers of this sort offered than the time-table will comfortably hold. Sometimes the only reason for such papers is that the author needs a pretext to get his travelling expenses paid by his employers; but that is not the full explanation. I believe that the primary motive is to be noticed as a serious contributor to the consensus, to be seen by the leaders, to act the part of membership in the College.

A scientific conference, as the venue of the face-to-face social interaction that governs an Invisible College, is thus a fascinating phenomenon, full of hidden meaning and symbolic ritual. Yet it would be a great mistake to ignore its genuine function as a place for the exchange of scientific information. The actual papers themselves may not be of great importance, but the informal discussions, the talk over lunch or at the bar, questions from the floor and the remarks of the chairman of the session, are the means by which the current consensus is dramatized to the participants. As I have already indicated, it is not at all easy to decide, just by reading the official literature, what is accepted as genuinely proven at any given moment in a fast developing science. One goes to a conference, not so much to hear in advance about some large discovery, or to pick up titbits of useful knowledge, but because this is the chance to talk to one's colleagues, to get their informal opinions on many doubtful questions—opinions which they might hesitate to put into print—and to hear from the leading authorities their views on the 'present state of the art'.

The very word 'authority' seems out of place in science, where, so we are told, the only arbiter of truth is Nature, interrogated by experiment. Yet in practice we defer enormously to the opinions of the acknowledged experts, the Senior Fellows of our Invisible College, the chairmen of conference sessions, the 'invited' speakers, the pundits who give courses at summer schools, write review articles, adjudicate between conflicting referees, edit journals, sit on all the grant-awarding committees, and generally seem to run Science for us. How can this élitism and autocracy be compatible with the essential modesty required of the man of Science, and the essential democracy and equality that ought to reign in the relations between scientist and scientist?

The answer is that Science is not that community of simple saints pictured in some of our hagiographies. It is a social system in which leadership and intellectual initiative naturally gravitate to the individuals who can exercise it. It is a commonplace, in the discussion of the concept of 'equality' in ordinary society, to remark that all men are not born as brilliant as Einstein, and therefore there is no reason why all men should be treated the same, etc., etc. The fact is that all theoretical physicists are not born as brilliant as Einstein, either, nor all biophysicists as smart as Crick, nor all radio astronomers as inventive as Ryle, and so on. The actual talents of those actively engaged in modern science spread over an enormous range, from the genius with 100 brilliant discoveries to his credit to the dullard with a Ph.D. and one uninteresting published paper in an obscure journal. One would not be human if one did not listen rather more carefully to the former than to the latter, nor if one did not seek to be heard at least in proportion to one's estimate of one's own standing on this scale.

It is not surprising, therefore, that an Invisible College devotes much of its hidden activity to the assessment of the relative standing of its members. A formal list, like that of the Politbureau in order of seniority, would, of course, break all the conventions of courtesy and social restraint; but there is always much discussion of individuals, and reputations are at the mercy of continual gossip in every laboratory around the globe. It is scarcely surprising that some eminent scientists become almost paranoic about protocol, and are capable of sensing a deliberate slight in the failure to be invited to a conference, or in the position of their lecture on the timetable, or in the number of lectures allotted to them in the summer school syllabus, or whatever it is that they judge to be symbolic of their standing in this little world. Their typical vanity is to attend more conferences than they can grasp the significance of, to speak longer and more often than they are allowed on the programme, to give the appearance of the repute to which they aspire.

Actually, the direct competitiveness of modern research has, I think, been somewhat exaggerated by sensational journalists. There are elements of direct rivalry for esteem, or for a coveted appointment, but the courtesies are usually preserved, and the other chap very very seldom cheats to win. Occasionally one finds oneself rather emotionally engaged in a controversy upon a scientific matter; but this, after all, is one's job, and one must learn to advocate powerfully, but to concede defeat gracefully, without making a personal issue of it. In a sense, a well-fought controversy between two spirited champions is a form of cooperation; like two barristers contesting a legal issue, they remain 'officers of the Court' with the common goal of seeing that justice is done, and are to be found dining amicably together that very evening, once they have doffed wig and gown.

Is this too idyllic a view of scientific controversy? It is true that many able scientists are flamboyant, witty men, capable of wounding more deeply than they intend in open debate. Others are, by nature, sensitive and jealous, easily hurt, and driven to malicious fury by what they regard as slights on their professional reputations. An idea that one has nursed for years, until it has become the apple of one's eye, becomes, when published to the world, an emotional attachment like one's own child. The insensibility of the scientific community to one's contribution seems like deliberate neglect or rejection, without reasons given to which one might reply.

What seems like a purely scholarly controversy may also hide more material conflicts, such as the claim to power or financial support. It is possible for the institutions of the protagonists to become involved, so that whole groups may feel that they must confront one another. Or there may be large questions of technical applicability at stake, as in the classic struggle between Oppenheimer and Teller over the feasibility of making a Hydrogen Bomb.

Yet the norms of ordinary Science strongly discountenance the *argumentum ad hominem*, the vitriolic personal attack, the public mud-slinging that is quite commonly observed amongst historians, philosophers, theologians and other quite respectable scholars. A genuine scientific controversy ought in principle to be about a question that can eventually be decided by appeal to experiment or better calculation. To become too deeply involved is therefore to take a serious risk of being proved quite categorically wrong. The commitment to the consensus principle advises the participants not to press their differences to a breaking point, which will be harmful to the whole enterprise.*

What makes so many scientists anxious and obsessed about their work is seldom mere personal rivalry, nor the fear of being anticipated or done down by unscrupulous competitors. There are plenty of prizes by way of useful discoveries to be made, and one only has oneself to blame if one deliberately challenges an opponent in a narrow field. The difficulty is simply to keep up with the ever-moving frontier, so that one's work acquires and maintains significance to others. After all, Nature is very tenacious of her secrets, and it always turns out to have been easier to get things wrong than right; at the same time, one is having to keep in the vanguard along with a larger and larger field of fellow scientists, all pressing on in the same direction. If one has, indeed, ridden for a while in the crest of a wave, it is disheartening to find

* The whole subject of the conduct of scientific disputes is discussed at length in a perceptive chapter by Hagstrom, who shows the strength of the social procedures used to isolate, neutralize or settle them. All I would add to this admirable account is the general point of this book—that the creation and preservation of a free consensus is the overriding aim of Science, and not a by-product of some other social or intellectual goal.

oneself left behind—or just washed up on the beach of an over-killed subject. The claim to speak with authority must continually be justified by further significant contributions; it is impossible to rest on one's oars, and coast along on reputation alone. The scholar does not seek material prizes for their own sake but as reassurance of his own continuing relevance to the scientific community and the advancement of knowledge

One might say, perhaps, that the rise and fall of authorities and the ceaseless reassessment of scientific reputations is a sort of personification of the continual critical revaluation of the intellectual attitudes for which they stand. The problem of creating a consensus of ideas becomes transformed into the creation of a consensus on the personal standing and credibility of the individuals who speak for these ideas. The tendency towards the cult of the hero, the divisive forces epitomized in the sectarian 'schools' of some non-scientific activities, are always a danger to Science. The continuous cool reassessment, by the younger generation and by the uncommitted sceptical masses, of the relative standing of the various conflicting authorities is the social manifestation of the consensus ideal at work.

Now, surely, it will be argued this sort of thing is illegitimate or unhealthy in Science, where all must be just humble seekers after truth. An attempt ought to be made to curb these excesses of individualism and to insist that we listen only to what is said, not who says it. This, I believe, is not merely psychologically unrealistic; it also conflicts with the consensus aim itself.

Scientific knowledge is not created solely by the piecemeal mining of discrete facts by uniformly accurate and reliable individual scientific investigators. The processes of criticism and evaluation, of analysis and synthesis, are essential to the whole system. It is impossible for each one of us to be continually aware of all that is going on around us, so that we can immediately decide the significance of every new paper that is published. The job of making such judgements must therefore be delegated to the best and wisest amongst us, who speak, not with their own personal voices, but on behalf of the whole community of Science. Anarchy is as much a danger in that community as in any tribe or nation. It is impossible for the consensus—public knowledge—to be voiced at all, unless it is channelled through the minds of selected persons, and restated in their words for all to hear.

I have said something in the previous chapter of the way in which such utterances are actually made—as referees' reports, review articles, books, lecture notes, etc. But we must still be concerned with the means by which they are also mediated and prevented from lapsing into dogma. As I have already suggested, the absence of formal grades of authority within the Invisible Colleges is a safeguard, and suggests that one should beware of the creation of such permanent labels by more worldly institutions, such as National Academies. This is the reason why, for example, the Royal Society must be extraordinarily careful in its choice of Fellows; it is not merely a matter of fairness to individuals in the award of a highly competitive and deeply felt personal honour; it also has its effects right through the intellectual structure of Science by giving explicit and permanent weight to the views of those who are thus distinguished. It could be argued—although this is a subtle matter which I forbear to pursue—that a more 'democratic' or 'egalitarian' system, in which any respectable scholar might expect, at the appropriate age, to achieve a certain official standing—as for example in the American academic world—is more in keeping with the fluidity of the assessment of authority implied by the consensus ideal.

It must also be re-emphasized that no field of Science nowadays is hermetically sealed from its neighbours. The spread from Pure Mathematics to Sociology has few major gaps, however imperfect our knowledge may be in many important regions. Because the Invisible Colleges are not 'organized' on any rational plan, they overlap one another in a multidimensional array, so that many scholars find themselves at the junctions of two, three or more 'fields'. As Polanyi has pointed out,* the validation of authority

* *Personal Knowledge* (p. 216) and in a powerful essay that strengthens much of the argument of this chapter, 'The Republic of Science', *Minerva*, 1, 54 (1962).

in any field is mediated by the opinions of authorities in neighbouring fields who are competent to make such judgements by the proximity of their intellectual interest. The whole battle line of Science is thus mutually self-supporting, and does not allow serious gaps to occur across its front. It is true that this system is liable to oscillate as a whole, as fashionable techniques are propagated from field to field, but it is relatively immune from petty tyranny and charlatanism. Thus, for example, if Lysenko had not been protected by State power, and had captured Soviet Genetics by the persuasiveness of his arguments, he would probably have been eliminated by the orthodox forces of the neighbouring sciences of Botany, Agriculture, etc., where his critics would have taken shelter.

This discussion of the nature of scientific authority brings out the importance of general freedom of speech and publication. It is quite evident that the natural forces tending to orthodoxy and subsequently to dogmatism are not absent from the scientific world; they are only held in leash by imaginative, critical, intellectual innovation and unorthodoxy. The cult of Creativity is healthy, so long as it is not allowed to overwhelm its twin opposing deity, Criticism. There are elements of the Ecclesia in Science—a priesthood in white coats, tending their gleaming glass and steel altars in cathedral-like laboratories, under the rule of abbots, bishops and cardinals. The analogy may seem crude and fanciful—but no more odd than that the Primitive Church became an organization that killed men in the name of the God of Love. We ignore the social nature of Science at our peril.

* * *

Re-reading what I have written in the last chapter or so, I find myself wondering whether I have presented too conservative a view of Science. Do the views expressed here imply that the scientific method, the scientific life and scientific community were at their best some thirty years ago (before my own time in fact), and that subsequent developments are leading into an epoch of decadence and decay, where the very spirit of the enterprise is being destroyed by opulence and the corruption of power?

The arrogance and spiritual insensitivity of many of those who write and talk in public about Big Science certainly tempts one into such a position. A mixture of politics with academic conceit has seldom looked pretty, and an extra component of technological brutalism is no improvement. Until recently Science has, perforce, had the virtues of poverty; now it must preserve itself from the vices of wealth. It would be easy to cry for a return to Little Science, to the peace and isolation of the Ivory Tower.

But the Arrow of Time knows its direction, and cannot be reversed. From Bacon onwards, the advocates of Science have supported the ideology of Progress, and have felt complete confidence that the Future was theirs. If I have seemed to be praising past ways, this is to some extent an effort to think 'countercyclically' (in Riesman's phrase) and to show what those ways were, before their purposes are quite forgotten.

The rapid taming and bureaucratization of Science—part perhaps of the same transformation of all our social institutions—is a major fact of our day. Each pronouncement on the relationship between Science and Government, each new Ministry, or Select Committee, or Presidential Advisory Committee, or Bureau, or Agency, of Science or Education or Technology, each new record percentage of research expenditure in the national income, each science-based industry and research corporation, implies further growth, 'rationalization' or 're-orientation' of institutionalized Science. It means more Science—which many scientists welcome—but it also means more organization of Science and scientists.

I do not wish to discuss the rationale of this development—such questions as whether all research can eventually be justified, or what proportion of the gross national product ought to be spent on building particle accelerators, and so on. The answers to such questions do not lie within Science itself, but imply a larger ethical and political framework of principles and values. But I hope that the general argument of this book may help to suggest some of the dangers and difficulties that could occur, within the scientific community, and with consequences for the value and validity of its products, by such a transformation of its social setting.

It is important, for example, that the formal institutions should be consistent with the informal social relationships discussed in this chapter. Scientists are not a leaderless mob; their work, as I have indicated, embeds them in a matrix of social obligations, invisible yet collegiate. To make the most of their efforts, and to keep them happy in their jobs, the lines of formal leadership and authority should coincide, as far as possible, with the informal relations. To put, say, an admiral in charge of a laboratory—even an admiral with a Ph.D.—might be as offensive and stupid as putting a professor (even a professor with a master's ticket) in command of a battleship. Whether or not professional scientists should be given executive power in general government, there is no doubt that the management of scientific research must be in the hands of those who are respected for their scientific and intellectual achievements.

It does not follow that managerial power should be thought of as the automatic 'reward' for scholarly prowess. To the scientist dedicated to his Invisible College and the pursuit of Knowledge this is not the goal of his ambitions. The British academic custom of making each professor the permanent Head of a Department is objectionable, not only for this reason but from the quite common circumstance that his scholarly authority may wane with age, leaving him only an arbitrary temporal power.

Indeed, power—moral force backed by a social organization—must necessarily be antipathetic to the sort of personal intellectual authority which is the essence of Science. The principles of free speech and equality between old and young, senior and junior, are officially obeyed in scholarly affairs, however much deference there may be in the informal social structure. A large-scale bureaucratic organization must, on the other hand, have a hierachical structure; it is essential to minimize the effects of this on its intellectual business.

For example, a system of committees, consultations and discussions of all substantial matters is essential. Those who decry the inefficiencies and delays of academic life, with its continual time-consuming committees, must remember this. The alternative could only be the sort of arbitrary power which earns for the 'Administration' of many American universities the nickname 'The Kremlin'.

Perhaps the most effective principles are those of decentralization and competition between parallel organizations. Ben-David has attributed the superiority of the nineteenth-century German universities to these factors which carried them far ahead of the 'orderly', over-centralized French system. The tendency towards orthodoxy and the suppression of unseemly criticism is quite strong enough in the informal institutions of the scientific community; it is very dangerous to give them official sanction by putting, say, the powers of appointment and promotion into the hands of a single group. As I have indicated, there is nothing like an exact consensus on the relative abilities of various scholars; a fragmentation of power into many separate institutions is essential if the full variety of opinions and tastes is to be allowed to show itself, and indeed if the critical and creative springs of Science are to continue to flow.

But the question we must ask is whether, in some way, the informal institutions, the ethics, the philosophy of Science can be strengthened to meet these attacks. The sad history of the movement amongst the Atomic Scientists against the use of nuclear weapons is not very encouraging. There scarcely exist any reputable quasi-political organizations for the defence of Science itself. National Academies and scientific societies have been very slow indeed to express their interest in such matters.

One might have hoped that the home of such defence would have been in the universities, which exist, after all, as quasi-independent corporations of scholars, all devoted, in their ways, to the pursuit of knowledge. But again, the history of the McCarthy era in the United States is not very encouraging in this respect. The modern scholar, in almost every discipline, tends to give his first allegiance to his Invisible College. In a very large modern university each department may be big enough to give the illusion that it subsists unto itself. Specialization and departmentalization have destroyed the general unity of scholarship at the very moment when interdisciplinary subjects are filling the intellectual gaps between the older Faculties. The very suggestion that I have made near the beginning of this essay, that there are perfectly reputable 'scientific' aspects of the study of History, may seem absurd to both historians and scientists, who have become used to thinking of each other as entirely separate species. An epoch of intellectual unification is going to be needed before the academic community can act as a group on any issue other than salaries and car-parking.

This is the reason why I regard the general subject matter of this book to be of the first importance. Even if my basic theme and its applications are unacceptable, even if I have got it all wrong, I hope that a discussion of the character of Science itself may help the proper incorporation of scientific knowledge and scientific research into the philosophy and life of our modern society. The notion of Public Knowledge raises subtle and complex issues of philosophy, psychology and sociology, comparable with the spiritual, ecclesiastical and political issues raised by the notion of personal responsibility in medieval theology. I can only hope that by indicating these issues, and giving examples of how the consensus ideal can help to resolve them, I may interest other scholars—scientists and humanists—in the grave problems that underlie them. Perhaps also I may have shown that Science itself is by no means as inhuman as it is sometimes painted.

8

SUMMING UP

'Anyone who thinks for himself exclusively and is consequently in a perpetual state of belief, i.e. of confidence in his own ideas, will naturally not trouble himself about the reason and motives which have guided his reasoning process. Only under the pressure of argument and opposition will he seek to justify himself in the eyes of others and thus acquire the habit of watching himself think, i.e. of constantly detecting the motives which are guiding him in the direction he is pursuing.' PIAGET

A scientific paper ends, by convention, with a few paragraphs entitled 'Conclusions'. To attempt this for a whole book is often to embark upon yet another round of pleas and rebuttals—or to risk exposing one's argument in all its naked triviality.

Yet the subject of this essay presents so many beguiling openings into byways of irrelevancy that it may help the reader if I try to express, in general terms, a view of the whole. In order to justify the promise made, in the first chapter, of indicating the power of the consensus principle, I have had to explain a number of features of scientific research, and of the scholarly life, that are not, perhaps, very familiar to the layman. For this exposition of a number of very important intellectual, psychological and social issues I make no apology; this is a subject where myth and mystery often take the place of personal knowledge, observation and experience. Nevertheless, I think we should try to return to the general principle at the end. What, after all, is being claimed in these pages?

There is not much doubt that, in a purely factual sense, Science is a form of Public Knowledge. The whole procedure of publication and citation, the abhorrence of secrecy, the libraries full of periodicals and treatises, *Lernfreiheit* and *Lehrfreiheit*—freedom to learn and freedom to teach—cosmopolitanism and internationalism, conferences, abstract journals and encyclopedias—all are in the service of the mutual exchange of information. If there is, indeed, a technical crisis in these procedures, if, as many people believe, the highways can no longer carry the traffic of ideas that pour into them, then Science itself will be severely hampered.

But merely to point this out tells us little about the nature of Science itself. I want to go further, and suggest that the absolute need to communicate one's findings, and to make them acceptable to other people, determines their intellectual form. Objectivity and logical rationality, the supreme characteristics of the Scientific Attitude, are meaningless for the isolated individual; they imply a strong social context, and the sharing of experience and opinion.

Here, perhaps, is a point beyond which many philosophically minded persons will not go along with me. The standard rationale of Science, the positivist ethic, the intellectual apparatus of 'The Scientific Method', the vast literature on inductive generalization, empiricism, etc., these all claim a somewhat more absolute validity than my argument will allow. I do not, of course, deny that Science advances by these means, nor am I suggesting that there is some other proof procedure, or language of argument that might somehow be revealed to us. To put it bluntly, let us have no truck with mystical ineffabilities of the Teilhard de Chardin ilk. On the contrary, I am arguing that all genuine scientific procedures of thought and argument are essentially the same as those of everyday life, and that their apparent formality and supposed rigour is a result of specialization. The demands of public communication, the pressures of overt criticism and comparison, have sharpened and strengthened these procedures so that we come to believe in them as authorities in their own right. Again, I do not suggest that scientific knowledge is not well-founded—but so is our ordinary knowledge of tables and chairs, men and marmosets, love and hate.

These procedures are, in fact, much less formalized than the conventional philosophical discussion would have us believe. Much of the analytical justification of Scientific Method that one finds in the books refers to very special aspects of scientific knowledge, particularly the grand theories that have, from time to time, erupted into the intellectual landscape. As Kuhn has so clearly said, these are not 'normal' science; they are profoundly disturbing and revolutionary, and yet in hind-sight, irresistible. Their very power to explain presents us with an infinity of different *ex post facto* justifications, many of which can be found to conform to our prescribed procedures. But most science, in practice, is not like this. Our fumblings are less systematic, our erroneous interpretations grosser —and our successful insight more penetrating—than can be catalogued in this way.

Moreover, young scientists do not learn to do research by studying books about Scientific Method. They use their native wits, and they imitate their elders. In the formative years of the doctorate, they acquire intellectual attitudes, technical procedures, and social conventions that fit them for membership of the scientific community. The ambition to excel in originality of thought is tempered with criticism; the orthodoxies of scientific education are broken through, and imagination is allowed space to roam. The graduate student not only learns the advanced technique of his subject and makes some small contribution to it; he become acquainted with the rules of scientific communication and controversy, and acquires his own internal version of the standards of argument and proof demanded by the scholarly world.

These rules, conventions and critical standards are, in my opinion, dominated by a single principle—their goal is the establishment and extension of a free intellectual consensus. Scientific standards of proof and criticism are harsh, because scientific knowledge must eventually be convincing, beyond reasonable doubt. Scientific communications must be refereed, to be sure that the published primary material of science is at least minimally plausible. Treatises and review articles are important because they give explicit voice to the current consensus. Seniority and 'authority' in Science represent the delegated power to speak for the scientific community, especially as to the apparent validity of proposed contributions to consensus opinion. Controversy is muted, so as to avoid the dangers of sectarianism, which represents the ultimate breakdown and failure of the consensus principle. And so on.

At this point we run into several difficulties. In the first place, we find that we cannot define the community over which the consensus is to be established, except rather narrowly as 'those educated and expert in the field'. This seems to suggest that Science may be in the nature of a conspiracy, or at best a delusion maintained within a self-penetrating circle: it seems to lose its universality. To this I can only answer that this is not an objection to my general argument, but is a genuine danger into which Science can, and sometimes does, run. Our only protection is the size of the scientific community and the interlocking of the various fields, so that new critical attitudes, expertise and ideas can always percolate from one discipline to another to correct local errors. It is obvious, however, that a general attitude of rational scepticism, and freedom to speak and write, are essential to the health of Science as a whole.

Another difficulty, which I have not discussed previously, is the actual status of the consensus principle. In suggesting that one can understand and explain many features of scientific life and thought in terms of this principle, I may be guilty of introducing a 'hidden variable', which is seldom a conscious motive of scienttists themselves. Here I may plead that I have found this point of view stated fairly openly in the writings of a number of serious commentators on the intellectual, psychological and social aspects of Science, although no one has previously attempted to bring together these aspects, and to work out its implications in all three dimensions simultaneously. I certainly do not hold it to be a metaphysical force, like a Hegelian 'Spirit of History', or the Marxian 'Dialectic'. Nor is it, apparently, some idealized, half forgotten constitutional doctrine, established in a Golden Age, like the Social Contract. The nearest analogy I can come to is the principle of Utilitarianism 'The greatest good of the greatest number', grasped as a criterion for conduct in a freely cooperating community. In other words, each individual scientist is to be seen as concerned mainly with putting forward his own ideas, trying to make discoveries for himself and therefore explicitly describing his thoughts and behaviour in essentially personal terms. But because he is, indeed, a member of the scientific community, because he is bound to communicate his ideas and make them public, he unconsciously makes allowances for the rational behaviour of others, and learns to put himself in their place.* He necessarily acts, therefore, both as author and critic, so as to maximize the area of agreement that he can achieve with others and hence, eventually, to make his own contributions acceptable. That this form of social control is imperfect is shown by many cases where crankiness, sectarianism and other pathological symptoms have developed in Science. One of my aims in this book has been to spell out the rationale of this form of enlightened self-interest, so that scientists themselves may become more aware of, and more responsive to, the necessities of their cooperative endeavour.

Finally, as I read popular journals and books about 'Science', I cannot help being disturbed by the use of the word to cover almost any conscious rational technique. My own experience being in the 'purest' of Natural Sciences—Theoretical Physics—I have automatically drawn my examples and analogies from this and neighbouring fields, and have argued that there is an essential difference between this sort of activity and Technology. Is it right, as I have suggested, to classify Sociology, Economics, even parts of History, with Physics, and then to draw a line of demarcation at the boundaries of practical Engineering and Medicine? To the vulgar eye, it must seem perverse to distinguish between a radio telescope and a radar set, or to enquire closely into the use being made of an electro-encephalograph before deciding in which pigeon hole to place its results. On this point I can only claim that one can make this distinction in terms of the consensus principle, and that it is still useful to have a name for all those activities which belong together in the way that I have indicated. The subject of 'rational technique' is even vaster and more diffuse than Science as I have delimited it, and calls into question the whole of our way of life, our ethical and moral standards, and the poetical quality of our earthly existence.

* Piaget makes a point of just this phenomenon in the growing child, as the origin of logical thought—see his *Judgement and Reasoning in the Child* (London: Routledge, 1928).

INDEX

Abstract journals, 119, 143
Academicism, 16, 31, 66, 73, 76, 79
Academies, national, 82, 105, 137, 141
'Acknowledgements', 100
Administration, 96, 101, 127–9, 140
Agriculture, 44
Algebra, 8, 26, 45, 46
Amateurism, 83, 87, 94
Anatomy, 68
Anthropology, 42, 49, 131
Archimedes, 4
Art of Scientific Investigation, The, 33
Arts, 1, 10, 16, 66, 101; *see also* Humanities
Associations, *see* Institutions, Societies
Astronomy, 4
Atomic Scientists, 141
Authority, 72, 113, 131–40, 145
Autocracy, 133

Bacon, Francis, 4, 88, 102
Barber, B., ix, 95, 96
Ben-David, J., 84, 141
Bernal, J. D., 125
Beveridge, W. I., 33
Biochemistry, 65, 67
Bohr, Neils, 114
Books, 89, 102, 104, 124
Braithwaite, R. B., xii
'Breakthrough', 52, 53
Britain
  Science in, 87, 137
  universities, 84, 140
Browne, Sir Thomas, 77
Buchdahl, G., xii
Bureaucracy 140

Caloric, 53
Cambridge, 4, 87, 105
Campbell, N., 30
Carnot, 60
Carroll, Lewis, 127
Cavendish Laboratory, 89
Cellular Biology, 53; *see also* Molecular Biology
Chemistry, 54, 65, 67
China, Science in, 10, 22
Citation Indexes, 59
Citations, *see* References
Civil Service, 84, 127, 128
Classicism, 16, 52, 74
Classics, 67, 85
Collaboration, 100, 131
Colloquia, 131
Committees, 140
Communality, 96
Communication, 102–26
  crisis of, 10, 125, 144
  formal, 109, 111, 116, 132–4
  informal, 104, 108–10, 132–4
  new techniques of, 47, 106, 108, 120, 125
Community, scientific
  common metaphysic of, 39, 83
  formal institutions of, 105, 117, 129, 132, 140
  fragility of, 88, 141
  freedom of, 39, 116
  growth of, 139, 146
  informal institutions of, 88, 108, 133, 140
  membership of, 63–5, 114, 145
  norms of, 77, 94–101, 115, 145
  openness of, 64, 93
Competition, 126, 134, 137, 141
Computers, 8, 13, 47, 120
Concept of the Positron, The, 50
Conferences, 93, 108, 109, 110, 132, 143
Consensible knowledge, 11, 39
Consensus, 9, 143, 145–7
  accepted by scientists, 73, 78, 116
  basis of scientific community, 96, 101, 132, 136, 142
  basis of scientific method, 30–2, 34–6, 38–9, 146
  changes continuously, 75, 79, 95, 98, 121–4
  contributors to, 63, 93
  features of, 43–4, 66
  lacking in non-Science, 15, 19, 22, 27, 29

Consensus (*cont.*)
origins of, 67–8
orthodoxy of, 53, 57, 69, 70, 97, 114, 122, 135, 137
publication of, 103, 109, 113, 118, 120, 122, 124, 126
vagueness of, 68, 123
Conservatism, 52, 54, 57, 122, 138
Consumers Association, 25, 27
Continental Drift, 56, 66, 75, 97 123
Controversy
scientific, 83, 84, 95, 100, 112, 115, 134, 145
non-scientific, 19, 28, 70
Copernicus, 53
Cosmology, 3, 18, 79, 114, 115
Courtesy, 58, 123, 133, 134
Cranks, 51, 56, 58, 64, 82, 97, 114, 147
Creativity, 12, 79, 81, 138; *see also* Imagination
Crick, F. H. C., 133
Criticism, Literary, 16, 70
Criticism, scientific
errors corrected by, 56, 66, 72
freedom of, 39
imagination and, 80, 114, 116, 138, 145
internalized, 34, 84, 86, 97, 115
publication of, 59, 83, 115, 121
referees and, 111–16
standards of, 79, 84, 87, 90
Crystallography, 67
Cytology, 67

Darwin, Charles, 70, 78
Davy, Humphry, 86
Dedication, 128, 130
Democracy, 133, 137, 140
Diagrams, 46, 118
Discovery, 48–52, 94–5, 99, 102, 124
'Discussion', 110
Dissertation, 60, 84, 89–90
Doctoral degree, 7, 60, 64, 84, 89, 117, 131
Dogmatism, 65, 66, 70, 75, 82, 116, 137, 138

Economics, 26, 68, 147
Editors, 86, 105, 112, 133
Education, scientific, 63–76, 94, 101, 145
dogmatism of, 69, 74, 75, 101
graduate, 72, 89–90
technological, 68, 86
undergraduate, 66–7, 81
Einstein, A., ix, 4, 65, 70, 114, 133
Elementary Particles, Physics of, 5, 108, 114, 123
Élitism, 133
Empiricism, 35, 37
Encyclopedias, 104, 143
Engineering, 23, 26, 68, 147
Error, 5–6, 8, 55–7, 66, 72, 90, 92, 121
Euclid, 4, 22, 38
Evidence, 14, 17, 20, 21, 35, 122
Evolution, Theory of, 53, 65, 78
Examiners, 89
Experience and Theory, 38
Experiment, 32–6, 40, 48–51, 102, 121
Experimental Method, 4

Falsification
of data, 35
of theories, 49, 53, 122
Faraday, Michael, 86
Fashion, 17, 53, 60, 62, 80, 89, 91, 99, 126
Feuer, L., 82
Formulae, 45, 47, 123
France, Science in, 141
Franklin, B., 82
Fraud, 35, 97, 134
Freedom, 38, 111, 113, 116, 138, 143
Freud, Sigmund, 28, 85, 114
From Max Weber, 85

Galileo, 32, 78
Garvey, W. D., 107
Genetics, 60, 67, 108
Geology, 4, 56, 66
Geophysics, 66, 97
Germany, Science in, 83–6, 88, 92, 141
Gilbert, William, 82
Greek Science, 10, 22
Griffith, B. C., 107
Gutenberg Galaxy, The, 45

Hagstrom, W. O., x, 95, 96, 99, 101, 131, 135

Hanson, N. R., v, 50
Hesse, M., xii
Hirsch, W., ix
History, 17–19, 28, 67, 70, 85, 142, 147
Humanities, 16, 20, 66, 70, 72, 102
von Humboldt, Baron, 84
Humility, 96

Idea of Justice and the Problem of Argument, The, 38
Imagination, 71, 75, 79, 86, 116
Impersonality, 78, 109, 118
India, Science in, 22, 109
Induction, 4, 30, 35
Information retrieval, 10, 120, 125
Input-Output Analysis, 26
Institutions, scientific, 88, 92, 117, 127–33, 135, 137, 140–2
Instruments of Communication, xi, 12, 42, 118
Intelligence, 81
Interdisciplinary subjects, 65, 114, 142
Internationalism, 93, 143
Intolerance, 28, 71
'Invisible Colleges', 61, 97, 108, 111, 130–3, 137, 141

Japan, Science in, 87–8
Jargon, 118
Jeffreys, H., 56, 97
Johnson, Samuel, 102
Jones, E., 85
Joule, J. P., 113
Journals, scientific, 104–19
  as social institutions, 117, 129, 143
  origins of, 104, 109, 114
  specialization of, 107, 113, 116, 119
  unrefereed, 110, 111, 115, 116
Judgement and Reasoning in the Child, 147

Kelvin, Lord, 43
Knowledge, 123
Körner, S., xii, 8, 38
Kornhauser, W., 128
Kuhn, T. S., xi, xii, 53, 54, 60, 65, 70, 72, 144

Lagrange, J. L., 46
Laslett, P., 43
Lavoisier, A. L., 53, 82
Law, 1–3, 13–16, 117, 127
Lectures, 89
Lernfreiheit and *Lehrfreiheit*, 143
'Letters', 104, 108, 110
Libraries, 102, 103, 119
Life of Sigmund Freud, The, 85
'Literature' of a subject, 58, 103, 120
Logic, 7, 8, 36–8, 41, 59, 86
Logic of Scientific Discovery, The, 49
Logico-inductive system, 5, 8, 9, 30
Lonely Crowd, The, 82
Lucas, J. R., xii
Lysenko, T. D., 116, 138

McCarthy, J., 141
McLuhan, H. M., 45, 46
Magic, 3
Management of Research, 25, 128
Mathematical argument, 8, 13, 36, 41, 45, 57, 75
Mathematics, Pure, 3, 4, 37, 86, 137
Matter, 3
Maxwell, J. C., 70
Mécanique Analytique, 46
Medicine, 23, 44, 68, 114, 117, 127, 147
Mendel, G. J., 60, 70
Meredith, P., xi, 12, 32, 38, 42, 118
Merton, R. K., 95, 96
Metaphysic of Science, 39, 66, 94, 96, 144
Meteorites, 14
Michelson-Morley Experiment, 48
Microbiology, 67
Milton, John, 13
Molecular Biology, 67, 98, 108
Monographs, 123, 125; *see also* Books, Treatises

National Institutes of Health, U.S., 111
Nature, 108
Nature
  Law of, 9, 15
  Philosophy of, 3, 22, 86, 101
Neurophysiology, 42
Newton, Isaac, 9, 42, 104, 105

Nobel Prize, 44, 98
Norms of scientific community, 94–101
Number, 43
Numeracy, 74

Objectivity, 78, 79, 92, 144
Observation, 35, 102
Occam's Razor, 125
Oppenheimer, J. R., 135
Originality, 95, 145
Orthodoxy, 52, 53, 56, 65, 89, 114, 138
Oxford, 87

Page charge, 117
Palaeontology, 17
Paper, scientific, 58, 103, 105, 109, 117, 118; 'contributed', 132; 'invited', 132
Paradigm, 70, 72; *see also* Consensus
Patronage, 131
Penicillin, 3
Perelman, C., 32, 38
Periodicals, *see* Journals
Personal Knowledge, xi, 8, 137
Pharmacology, 68
Ph.D., *see* Doctoral degree
Philosophy, 1, 2, 5, 22–3, 28, 31, 135
Phlogiston, 53
Photography, 46
Physics, 7, 22, 41–2, 44, 54, 69, 75; *see also* Elementary Particles, Physics of
  Theoretical, 3, 37, 147
Physiology, 7, 67, 68
Piaget, J., 33, 133, 147
'Piltdown Man', 35
Plagiarism, 104, 108, 135
Pocock, J. G. A., xii
Poetry, 1–2, 16, 20, 39, 147
Polanyi, M., xi, 8, 35, 38, 65, 116, 137
Political Theory, 22, 69
Popper, K., 49
Positivism, 5, 32, 55, 72, 75, 122, 144
Positron, 50
Post-doctoral Fellowships, 92
Prediction, 41, 51
Preprints, 108, 110
Price, D. de S., 61
Printing, 11, 45–6, 106
Priority of discovery, 94–5, 98, 107, 135
Probability Theory, 44
Professionalism, 83, 94, 117, 127
Progress, 139
Protestantism, 10, 19, 82
Psychoanalysis, 28, 114
Psychology, 3, 44, 74
Publication, 102–26. *See also* Journals, Abstract journals, Books, Review articles, etc.
  basic to Science, 39, 97, 103, 117, 143
  critical standards of, 55, 86, 111–16, 117
  exposition of consensus by, 73, 124, 125
  forms of, 45, 119, 120, 122
  freedom of, 111, 113
  historical growth of, 104, 114, 119
  new techniques of, 48, 106, 126
  pressure for, 85, 117
  style of writing in, 9, 34, 70, 84, 90, 109, 112
  uncriticized, 108, 110, 111, 116

Quantum Theory, 41, 53, 65, 89, 114, 124

Rayleigh, Lord, 113
Recognition, 96
Referees, 79, 86, 111–17, 145
'References', 58, 103, 122, 143
Relativity Theory, 4, 6, 40, 53, 65, 114
Religion, 1–3, 10, 21–2, 39
Reputation, 133, 136
Research, 98, 99, 129
  industrial, 25, 128
  training in, 72–6, 90–1
Retrodiction, 41
Review articles, 86, 95, 100, 107, 122, 133, 145
Revolutions, scientific, 54, 71, 108, 144
Rhetoric, 31–2, 41, 118
Riesman, D., 82, 129

Rose, J. A., x, xii
Royal Society, 61, 83, 87, 105, 113, 137
Rumford, Count, 53
Rutherford, Lord, 89
Ryle, M., 133

Scepticism, 32, 54, 91, 96, 122, 136
'Schools of Thought', 19, 28, 60, 70, 86, 136
Science, belief in, 38, 40
  corruption of, 100, 117, 138, 147
  ecclesiastical tendencies of, 138
  fragmentation of, 60–1, 65
  History of, 66
  metaphysic of, 39, 66, 94, 96, 144
  'normal', 60, 67, 71, 76, 115, 124, 145
  norms of, 94–101
  origins of, 10, 22, 39, 105
  phases of development, 51–4, 123
  philosophy of, 6, 31, 37, 66, 74, 144, 146
  psychological dimension of, 11–12, 38, 58, 72, 78, 101, 146
  as Public Knowledge, 8, 144, 147
  as self-validatory system, 114, 146
  sociology of, ix, 10, 94–101, 130, 146
  and the State, 116, 128, 130, 139
  revulsion from, 74
Science and the Modern World, 22
'Science of Science', 101
Science since Babylon, 61
Scientific attitude, 77–9, 85, 116, 144
Scientific Community, The, x, 95, 99, 101, 131, 135
Scientific Intellectual, The, 82
Scientific Method, 9, 30, 31, 37, 48, 66, 72, 104
Scientism, 72, 74
Scientists
  allegiances of, 25, 128, 141
  bureaucratization of, 127
  career problems of, 60, 97–9, 107, 109, 131, 134
  conservatism of, 52, 122
  courtesies between, 58, 122
  disinterestedness of, 96
  education of, 63–76
  honesty of, 34, 78, 97, 130
  humility of, 96
  inequality of, 133
  intelligence of, 81
  literary style of, 118
  management of and by, 25, 101, 129, 140
  numbers of, 94
  personalities of, 68, 74, 81, 82, 91, 134
  philosophical views of, 5, 6, 39, 61, 102, 124
  political views of, 78, 130
  professionalism of, 81, 94, 127
  psychology of, 79–82, 91, 97
  scepticism of, 54, 96, 122
  training of, 7, 77–101
  vanity of, 134
Scientists in Industry, 128
Secrecy, 97
Seminars, 89, 108, 131
Shils, E. A., xi, xii
Smoking and lung cancer, 19, 45
Social Function of Science, The, 125
Social Sciences, 26–8
Social System of Science, The, x, 96, 131
Societies, scientific, 117, 129
Sociology, 3, 16, 27, 44, 67, 137, 147
Sociology of Science, The, ix, 84, 95
Specialization, 61, 105, 118, 119, 124, 137, 142
Speculation, 79, 84, 98, 114
Statistical inference, 26, 27, 43–4
Stone, J., xii, 13
Storer, N. W., x, 96, 131
Structure of Scientific Revolutions, The, xi, 53, 65, 70
Student
  graduate, 72–4, 89–90, 100
  undergraduate, 71
Summer schools, 133, 134
Superconductivity, 53
Symbols, 45

Teaching, 71, 73–6, 90, 96, 127, 129, 143, 145
Team research, 91

Technology, 1–2, 10, 23–6, 44, 68, 91
Teilhard de Chardin, 144
Television, 47
Teller, E., 135
Textbooks, 70, 73, 95
Theology, 18, 22, 28, 39, 135
Theory, 36–8, 40–1, 48–9, 51–3
Thermodynamics, 53, 60
Todhunter, Isaac, 633
Torment of Secrecy, The, xi
Tradition, 88, 92, 100
Treatises, 86, 104, 123, 125, 145
Truth, 5, 6, 14, 28

Understanding Media, 45, 46
United States, Science in, 88–92, 137, 141
Universalism, 96, 146
Universities, 83, 84, 88, 127, 129, 137, 140, 141

Vanity, 134
Verification, 122
Vocation, 85, 86, 130

Weber, M., 30, 67, 85
Wegener, A. L., 56, 97
Weiner, C., xii
What is Science, 30
Whitehead, A. N., 22
Wissenschaft, 20, 85
World We Have Lost, The, 43

Young, R. M., xii

Zen poetry, 36, 78